SpringerBriefs in Computer Science

More information about this series at http://www.springer.com/series/10028

Jinbo Xu · Sheng Wang
Jianzhu Ma

Protein Homology Detection Through Alignment of Markov Random Fields

Using MRFalign

 Springer

Jinbo Xu
Toyota Technological Institute
Chicago, IL
USA

Jianzhu Ma
Toyota Technological Institute
Chicago, IL
USA

Sheng Wang
Toyota Technological Institute
Chicago, IL
USA

ISSN 2191-5768   ISSN 2191-5776 (electronic)
SpringerBriefs in Computer Science
ISBN 978-3-319-14913-4   ISBN 978-3-319-14914-1   (eBook)
DOI 10.1007/978-3-319-14914-1

Library of Congress Control Number: 2014960093

Springer Cham Heidelberg New York Dordrecht London

Springer International Publishing AG Switzerland is part of Springer Science+Business Media
(www.springer.com)

Preface

This short book is derived from our paper entitled "MRFalign: Protein Homology Detection Through Alignment of Markov Random Fields," which won the best paper award at a premier computational biology conference RECOMB2014 and also appeared at PLoS Computational Biology. The intended audience consists of students and researchers involved in developing computational methods for biological sequence analysis and protein structure and functional prediction, those who use sequence analysis tools to study biology problems, and those who would like to have a general idea about protein homology detection and fold recognition. We hope that the new Markov Random Fields (MRF) method described in this book will intrigue further study of protein homology detection and fold recognition. We also hope that the tool described in this book will be helpful for readers with biology backgrounds who need to quantify and analyze protein sequences to answer interesting questions.

This book covers sequence-based protein homology detection, a fundamental and challenging bioinformatics problem with a variety of real-world applications. The book first surveys a few popular homology detection methods such as the position-specific scoring matrix (PSSM) or Hidden Markov Model (HMM)-based methods and then is devoted to a novel MRF-based method which was recently developed by our group. Compared to HMM and PSSM, MRF can model long-range residue–residue interaction and thus, MRF-based methods are much more sensitive than HMM- and PSSM-based methods for remote homolog detection and fold recognition. This book also describes the installation, usage, and result interpretation of our programs implementing the MRF-based method.

The book is organized into four chapters. Chapter 1 describes the background and surveys the existing popular methods of homology detection and fold recognition. Chapter 2 describes a novel MRF-based method for homology detection and fold recognition. In particular, it covers how to build an MRF model for a protein sequence, how to score the similarity of two MRF models, and how to generate an MRF–MRF alignment optimizing the scoring function. Chapter 3 is devoted to the software implementing the ideas presented in Chap. 2, covering installation, usage, and result interpretation of the software. Chapter 4 describes the experimental results of our MRF-based method for homology detection and fold recognition. Finally, conclusions are drawn in the last part of the book.

We are indebted to a few Ph.D. students in our group such as Dr. Jian Peng (now a faculty member at UIUC Computer Science Department), Dr. Feng Zhao, and Mr. Zhiyong Wang. We are also thankful to Dr. Söding, who developed the popular HHpred program for homology detection. It is their previous excellent work that leads to the MRF-based method described in this book.

Contents

Chapter 1
Introduction

Abstract This chapter describes background and surveys existing popular methods on homology detection and fold recognition. In particular, this chapter reviews homology detection methods from the following perspectives: alignment-free versus alignment-based, sequence-based versus profile-based, and generative versus discriminative machine learning. Finally, this chapter also reviews a few popular scoring functions for sequence-based or profile-based protein alignment.

Keywords Homology detection · Fold recognition · Alignment-free homology detection · Alignment-based homology detection · Profile-based protein alignment

1.1 Background

High-throughput genome sequencing has been yielding a large number of biological sequences without accurate functional and structural annotations [1, 2]. Due to experimental complications and obstacles in structural and functional analysis, the gap between the number of available protein sequences and the number of proteins with experimentally determined structures and functions has greatly increased in recent years [3, 4]. As such, novel bioinformatics methods that can link proteins without annotations to their homologs with accurate annotations are needed. However, a large percentage of proteins have no solved structures, so it is important to unravel protein relationship using sequence information. Meanwhile, homology detection and fold recognition are two essential techniques used to detect if two proteins are homologous or share similar folds [5, 6].

Two proteins are said to be homologous if they share a common evolutionary origin. Sequence information is often used to infer if proteins are homologous or not and their structure and functional relationship. If two proteins share high sequence similarity, say above 40 % sequence identity [7, 8], they are very likely to be homologous and have similar structures and in many cases also similar functions. It is observed that proteins sharing low sequence identity may still be remotely homologous. Nevertheless, homology detection is very challenging when

© The Author(s) 2015
J. Xu et al., *Protein Homology Detection Through Alignment*
of Markov Random Fields, SpringerBriefs in Computer Science,
DOI 10.1007/978-3-319-14914-1_1

the proteins under study share <30 % sequence identity. The problem of detecting homologous proteins with relatively low (<30 %) sequence identity is called remote homology detection. Remote homology detection is related to protein fold recognition, which is to infer if proteins sharing low sequence identity have similar structural folds or not. In this book we do not explicitly distinguish remote homology detection from fold recognition since many remote homology detection methods including those presented in this book also apply to fold recognition.

In the following sections, we will describe the state-of-the-art methods of homology detection and fold recognition, which could be roughly grouped into two main categories: alignment-dependent and alignment-free methods. Sequence information is essential to homology detection and fold recognition. Nevertheless, some methods also use predicted structure information. In this book, we mainly focus on sequence-based, alignment-dependent methods for remote homology detection and fold recognition that detects remote homologs based on protein alignment using mainly sequence information.

1.2 Related Work

Protein homology detection and fold recognition have been extensively studied and good progress has been made. More than 5,000 research articles indexed in PubMed (http://www.ncbi.nlm.nih.gov/sites/entrez) show relevance to "fold recognition" or "remote homology detection". See Fariselli et al. [9], Wan and Xu [10], Lindahl and Elofsson [11], and Jones et al. [12] for reviews on some widely-used computational methods and tools for remote homology detection and fold recognition.

This section reviews existing methods for remote homology detection and fold recognition. We will describe alignment-free approaches such as kernel-based fold recognition methods that classify a protein sequence into a specific fold class without aligning proteins. We will also describe alignment-dependent approaches that conduct homology detection and fold recognition based upon protein alignments. All the methods employ a few basic information sources including protein amino acid sequence and sequence profile. Meanwhile, sequence profile encodes evolutionary information of a protein and is usually derived from multiple sequence alignment (MSA) of close sequence homologs. In addition, some methods also make use of predicted structure information or even native structure information when available. See Fig. 1.1 for the classification of existing homology detection methods.

1.3 Alignment-Free Methods for Homology Detection and Fold Recognition

Alignment-free methods for homology detection and fold recognition do not explicitly build protein alignments. In particular, alignment-free methods represent a protein sequence or profile as a feature vector and then identify proteins of similar

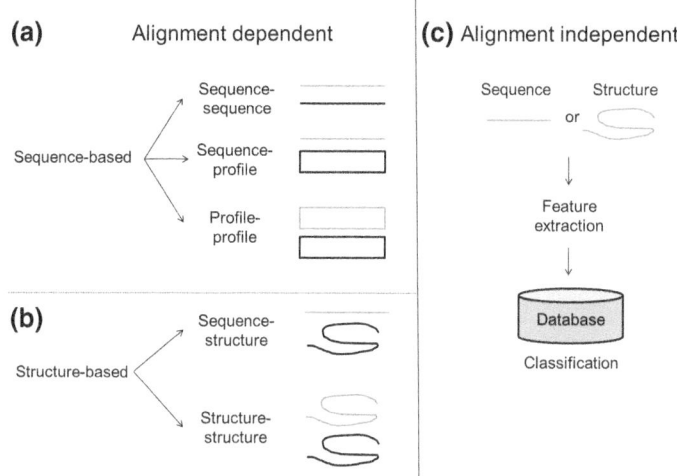

Fig. 1.1 Overview of homology detection methods. **a** Sequence-based, alignment-dependent methods. **b** Structure-assisted, alignment-dependent methods. **c** Alignment-free methods

feature vectors as homologs. Early methods such as [13, 14] use a straightforward method to compare feature vectors, so they are not very sensitive. The application of machine learning greatly improves alignment-free homology detection. Machine learning methods formulate homology detection and fold recognition as a classification problem. One typical way is to build a binary classifier for each specific protein class (e.g., superfamily and fold) to identify protein sequences belonging to this class. It was shown that given sufficient training data discriminative or supervised machine learning is in general superior to generative or unsupervised learning. In particular, a few discriminative learning methods have been developed using Support Vector Machines (SVM) [15–17]. The SVM methods primarily differ in their kernel functions used to measure the distance (or similarity) between two proteins. Some example SVM methods include SVM-Fisher [17], SVM-Pairwise [18], SVM with the spectrum kernel [19] and SVM with the mismatch kernel [20]. See [21] for a review of these methods. These SVM methods are reported to outperform the simple feature comparison methods [18, 19].

Alignment-free methods face two major challenges. One is how to represent a protein as a feature vector that contains enough information for homology detection. Many popular machine learning methods such as SVM require that a feature vector shall have a fixed dimension regardless of the length of a protein. This implies that position-specific protein features have to be compressed or transformed, which might lead to large information loss. The other issue is that a feature vector actually encodes information in a complete protein sequence, so it is challenging for alignment-free methods to recognize homologous domains in multi-domain proteins in the case that they are homologous only by one of their domains.

1.3.1 Generative and Discriminative Learning for Alignment-Free Homology Detection and Fold Recognition

Alignment-free methods represent a protein (sequence) as a feature vector so that homologs can be identified by directly comparing the resultant feature vectors. To compare feature vectors, alignment-free methods usually employ a machine learning method to classify feature vectors into classes. Two types of machine learning methods are employed: generative learning and discriminative learning [22]. Generative learning methods [23, 24] use a probabilistic model to define the occurring probability of a feature vector and automatically find patterns in the data. The main issue is that generative learning methods may not be sensitive enough to discriminate distantly-related proteins. It also needs lots of non-redundant data to estimate parameters for a generative model, which is not available for some protein families.

Discriminative learning overcomes these issues partially by directly training a machine learning model to differentiate homologous proteins (i.e., positive examples) from non-homologous proteins (i.e., negative examples). Existing discriminative learning methods mainly differ in feature representation and extraction schemes and the employed machine learning models. A few popular supervised machine learning methods have been explored including k-nearest neighbor [25], decision trees (random forests) [26, 27], neural networks [28, 29] and Support Vector Machine (i.e., kernel-based methods). It has been reported that tested on a few publicly available datasets Support Vector Machine (SVM) outperforms the others [30]. Besides SVM, probabilistic graphical models, such as Conditional Random Fields (CRF), are also used for fold recognition [31].

1.3.2 Kernel-Based Learning Methods for Alignment-Free Homology Detection

A few kernel-based methods have been developed for alignment-free homology detection and fold recognition. Their performance critically depends on protein features employed to model a protein sequence (or profile) and the employed kernel functions. Gaussian kernel functions are widely used and yield good performance. Ding and Dubchak [32] developed a multi-class Support Vector Machines (SVM) method for fold recognition that achieves an accuracy of 56 % on a dataset of 27 protein folds. The feature vector employed by this method mainly encodes the amino acid composition of a protein sequence. SVM-Fisher [17] represents a protein by a vector of Fisher scores extracted from profile Hidden Markov model (HMM) and employs SVM to classify profile HMMs based upon their feature vector. Shen and Chou [25] developed an ensemble classifier PFP-Pred, which uses a sequence feature called amphiphilic pseudo amino acid and also considers sequence-order information. PFP-Pred improves homology detection and fold recognition accuracy to 62 % on the test datasets used by Ding and Dubchak [32]. Dong et al. [33] developed a method called ACCFold with an overall accuracy of 70.1 %. ACCfold employs autocross-covariance (ACC) transformation to convert a PSI-BLAST sequence profile (i.e., position-specific scoring matrix) into a series of fixed-length vectors, which are then input to an SVM classifier for fold recognition. Ghanty and Pal [34] developed a fold recognition method that uses a bi-gram histogram to represent a protein sequence.

Recently Zakeri et al. [30] have developed a method called GEOMEAN and reported that GEOMEAN can achieve fold recognition success rate of above 80 %. GEOMEAN achieves this by taking a geometry inspired mean of different kernel matrices (functions) instead of using a linear combination. SVM-Pairwise [18] is another sensitive method that combines SVM and Pairwise protein alignment to achieve superior performance. SVM-Pairwise is not a strict alignment-free method since it uses a kernel function defined on protein sequence alignment. SVM-Pairwise was tested with both dynamic-programming-based alignment and BLAST alignment. SVM-Pairwise is among the best methods in terms of accuracy, but it is slow since it takes time to build alignments for large proteins. In addition, false positives in alignment (i.e., two unrelated residues are aligned) may impact homology detection rate of SVM-Pairwise.

1.4 Alignment-Based Methods for Homology Detection and Fold Recognition

Alignment-based methods detect homologs by first aligning a query protein to each of the subject proteins in the database and then rank and select homologs based upon alignment quality, which is evaluated by a scoring function. Alignment-based homology detection faces two major challenges. One is to design a good scoring function that can yield accurate protein alignments. The other is to select homologs based upon alignments, which is usually done by evaluating statistical significance (e.g., calculating E-value) of a raw alignment score or by machine learning.

According to information sources used for proteins under study, alignment-based methods can be grouped into three categories: sequence-sequence (i.e., primary sequence information for both proteins under comparison), sequence-profile (i.e., primary sequence information for one protein and sequence profile for the other) and profile-profile (i.e., sequence profiles for both proteins), as shown in Fig. 1.2. Generally speaking, sequence-sequence methods are less sensitive than sequence-profile methods, which in turn are less sensitive than profile-profile methods. However, sequence-sequence methods are more specific than sequence-profile methods, which in turn are more specific than profile-profile methods. See a review by Wan and Xu [10] for a list of sequence-sequence, sequence-profile, and profile-profile alignment methods developed in the past few years.

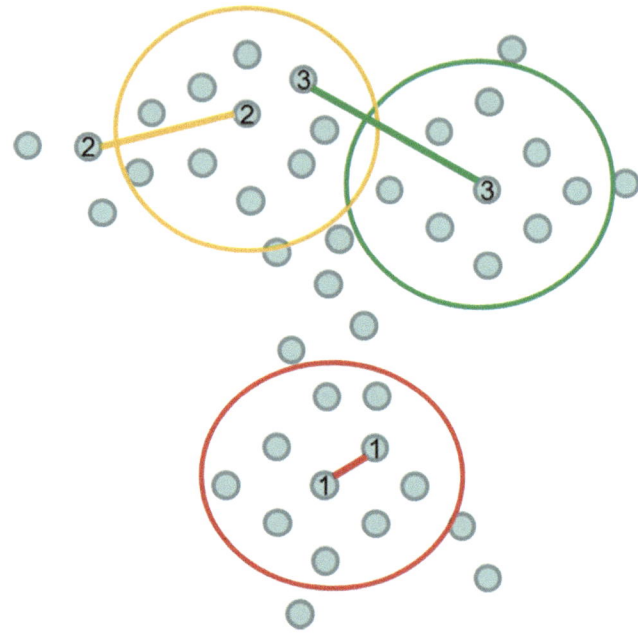

Fig. 1.2 Illustration of sequence-sequence, sequence-profile and profile-profile comparison methods for homology detection and fold recognition. A node represents a protein and the distance between two nodes represents their closeness. The *large circles* in *red*, *blue* and *green* indicate three different protein families with similar fold. In this figure, two proteins marked with *1* belong to the same protein family, so their homologous relationship can be detected through sequence-sequence comparison. Two proteins marked with *2* are not in the same protein family, but they are still evolutionary related and their homologous relationship can be recognized by sequence-profile comparison. Two proteins marked with *3* are distantly-related with similar folds, and their relationship may be recognized by profile-profile comparison

Sequence-sequence or pure sequence-based methods detect homologs by mainly aligning two primary sequences. They are good for close but not remote homology detection. Existing sequence-based methods mainly differ in alignment algorithms, amino acid mutation score and gap penalty. Some methods such as the Needleman-Wunsch [35] and Smith-Waterman algorithms [36] employ dynamic programming to build alignments, while others such as BLAST [37] and FASTA [38] use more efficient heuristic-based alignment algorithms. BLOSUM [39] and PAM [39] are two widely-used amino acid substitution matrices to score similarity of two aligned residues. An affine function is used to penalize gaps (i.e., unaligned residues) in an alignment.

Alignment-based homology detection can be improved by using evolutionary information such as PSI-BLAST sequence profile [37] or profile Hidden Markov Model (HMM) [6]. A few methods have been developed to align one primary sequence to one sequence profile or align two sequence profiles. For example, HMMER [40] and SAM [41] are two tools that align one primary sequence to one profile HMM. Other sequence-profile alignment tools include DIALIGN [42] and FPS [43]. HHpred [44], FORTE [45], and PICASSO [46] are some tools that use profile-profile alignment for homology detection. They have shown better performance than sequence-sequence or sequence-profile methods for remote homology detection.

In the following sections, we will present more details on the state-of-the-art methods for alignment-based homology detection and fold recognition.

1.4.1 Sequence Alignment for Homology Detection and Fold Recognition

Sequence alignment is a basic method for homology detection. The underlying reason is that two homologous proteins shall align well to each other (i.e., there are many conserved residues in their alignment and few gaps). As such, we can infer the relationship of two proteins by sequence alignment quality. A simple rule is that if the sequence identity of two proteins (i.e., the percentage of identical residues in their alignment) is high (e.g., >40 %), then it is very likely that they are homologous. The limitation of sequence alignment lies in that it cannot reliably detect homologous relationship when proteins under study are not very close to each other. In particular, homology detection by sequence alignment is not very reliable when the similarity of two proteins falls into the twilight zone, i.e., the sequence identity of two proteins is less than 25 %. However, in many cases two proteins sharing low sequence identity may still be homologous and share some important structural and functional properties.

The quality of one alignment cannot be accurately judged by sequence identity. Instead, a more sophisticated score is needed to quantify protein similarity based on the given alignment. A typical scoring function calculates a ratio between the likelihood of two proteins being homologous (or evolutionarily related) and that of being non-homologous (or evolutionarily unrelated). We can use two amino acid substitution models to estimate the probability of two proteins being homologous and non-homologous, respectively. The probability model for "non-homologous" is also called null model, describing the case that two aligned residues are evolutionarily unrelated. Let X and Y denote the amino acid types of two aligned residues. The null model calculates the occurring probability of X and Y being aligned as $(X = i, Y = j) = P_i P_j$, where P_i and P_j are the background probability of amino acid i and j, respectively. A few probability models such as PAM [39] and BLOSUM [39] have been developed to estimate how likely two aligned residues are evolutionarily related. PAM estimates the relatedness of two aligned residues starting from single point mutations. BLOSUM derives amino acid substitution model from blocks of multiple sequence alignment.

Let $P(X = i, Y = j | related) = P_{rel}(i, j)$ denote the probability of two aligned residues being evolutionarily related (i.e., one substitutes the other during evolution), then the amino acid substitution score or mutation potential between i and j is defined as follows.

$$M(i,j) = \log\left(\frac{p_{rel}(i,j)}{p_i p_j}\right) \tag{1.1}$$

The higher the $M(i,j)$, the more likely the two aligned residues are related. Summing up $M(i,j)$ over all aligned positions and deducting gap penalty yields a quality score of one alignment. $M(i,j)$ can also be interpreted in a different way. By Bayes' rules, we have the following equation.

$$\begin{aligned} P(related|X=i, Y=j) &= P(related)\frac{P(X=i, Y=j|related)}{P(X=i, Y=j)} \\ &= P(related)\frac{p_{rel}(i,j)}{p_i p_j} \end{aligned} \tag{1.2}$$

If we assume that the prior probability $P(related)$ is a constant, Eq. (1.2) implies that the log-likelihood of two aligned residues being evolutionarily related actually equals to their mutation potential plus a constant.

Given a scoring function, dynamic programming algorithm can be used to generate an alignment between two protein sequences, which guarantees to yield an alignment optimizing the scoring function. However, dynamic programming has computational time proportional to the length product of two proteins under alignment, which may be too slow in many applications of homology search especially when a large database of subject proteins is searched for homologs. To speed up, some heuristic methods, such as BLAST, are developed to generate suboptimal alignments and detect close homologs much more efficiently.

1.4.2 Profile-Based Alignment for Homology Detection and Fold Recognition

To improve remote homology detection, profile-based protein comparison is developed. The sequence profile of a protein encodes its evolutionary information and is built from a set of close homologs. That is, instead of aligning primary sequences, we may align/compare two protein sets, each containing close homologs of a protein in question. To make comparison or alignment easy, a set of protein homologs is usually represented as a sequence profile. The utilization of sequence profiles has increased the sensitivity of homology detection by three times over pure sequence comparison [47].

PSI-BLAST can be used to generate sequence profile of a protein. PSI-BLAST finds close homologs of this protein from a large sequence database such as the NCBI non-redundant (NR) database [48], build a multiple sequence alignment (MSA) of these homologs, and then convert the MSA to a sequence profile. For example,

PSI-BLAST represents the sequence profile as a position-specific frequency matrix (PSFM) or position-specific scoring matrix (PSSM), which is widely-used in many applications such as homology detection, fold recognition and protein structure prediction [45, 49, 50]. Both PSFM and PSSM have dimension of $20 \times N$, where N is the protein sequence length. Each column in a PSFM contains the occurring frequency of 20 amino acids at the corresponding sequence position. Accordingly, each column in a PSSM contains the potential of mutating to 20 amino acids at the corresponding position. A good sequence profile shall include as much information in the MSA as possible. In addition to representation, the quality of a sequence profile depends on the following factors: the number of PSI-BLAST iterations, the E-value cutoff used to determine if two proteins are homologous or not, and the sequence weighting scheme [39]. It also depends on how to include amino acid pseudo-counts in converting amino acid occurring frequency to mutation potential.

Profile Hidden Markov Model (HMM) [51] is another way to model an MSA of protein homologs. Profile HMM is better than PSFM/PSSM in that the former takes into consideration correlations between adjacent residues and also explicitly models gaps, so profile HMM on average is more sensitive than PSSM/PSFM for protein alignment and remote homology detection [40, 44]. In particular, a profile HMM usually contains three states: match, insert and delete. A 'match' state at an MSA column models the probability of residues being allowed in the column. It also contains emission probability of each amino acid type at this column. An 'insert' or 'delete' state at an MSA column allow for insertion of residues between that column and the next, or for deletion of residues. That is, a profile HMM has a position-dependent gap penalty. The penalty for an insertion or deletion depends on the HMM model parameters in each position. By contrast, traditional sequence alignment model uses a position-independent gap penalty. An insertion or deletion of x residues is typically scored with an affine gap penalty, say $a + b(x - 1)$ where a is the penalty for a gap opening and b for an extended gap.

A few popular homology detection programs such as HHpred [44] and HMMER [40] use profile HMMs for remote homology detection. Pfam [52, 53] and SUPERFAMILY [54] are two large publicly available libraries of profile HMMs of common protein domains. In many applications, profile HMM has demonstrated better performance than PSI-BLAST sequence profile (i.e., PSFM/PSSM) in terms of alignment accuracy and homology detection success rate [6, 44]. However, both profile HMM and PSFM/PSSM are restricted in that they cannot model long-range residue correlation in an MSA [55].

1.4.3 Scoring Function for Profile-Based Alignment and Homology Detection

A key component of profile-based homology detection method is the scoring function, which measures the similarity of one sequence and one sequence profile or that of two sequence profiles. Unlike protein sequence alignment that can use an amino acid substitution matrix such as BLOSUM62, profile-based alignment needs a different scoring function. Nevertheless, some scoring functions for primary sequence-based homology detection can be generalized to profile-based methods. In the following subsections, we briefly review the scoring functions for (1) the alignment between one primary sequence and one PSI-BLAST sequence profile (i.e., PSFM or PSSM), and (2) the alignment between two PSI-BLAST sequence profiles. A slight change of the scoring functions can apply to the case when a profile is represented as an HMM.

1.4.4 Scoring Function for Sequence-Profile Alignment and Comparison

Given a primary sequence and a sequence profile, to determine their similarity, one strategy is to estimate how likely the primary sequence is a sample from the probability distribution encoded by the sequence profile. Let α denote the probability distribution of 20 amino acids at a specific MSA or profile column. Supposing that this profile column is aligned to amino acid j in the other protein, to determine if amino acid j is a sample from α, the following score can be used.

$$score(\alpha, j) = \log \frac{\alpha_j}{p_j} \qquad (1.3)$$

where α_j is the probability of amino acid j at this specific profile column and p_j the background probability of j. That is, Eq. (1.3) calculates the log-odds ratio of amino acid j being observed at this specific profile column. The larger the score, the more likely amino acid j is generated from the distribution α instead of the background distribution. Summing up Eq. (1.3) over all aligned positions yields a score for the whole alignment between one primary sequence and one sequence profile. The larger the alignment score, the more likely that the primary sequence is a sample from the probability distribution encoded by the sequence profile. Therefore, the alignment score quantifies the similarity between the primary sequence and the sequence profile.

Another scoring method is to generalize the sequence-sequence scoring in Eq. (1.1) to sequence-profile scoring. Let $P(i \rightarrow j) = P_{rel}(i,j)/P_i$ denote the probability of amino acid i mutating to amino acid j, where $p_{rel}(i,j)$ is the probability of two amino acids i and j are evolutionarily related and p_i is the background probability of amino acid i. Summing up the 20 possible amino acids for i according to the probability distribution α, we can calculate the probability of j mutating from an amino acid distribution α as follows.

$$\sum_{i=1}^{20} \alpha_i P(i \rightarrow j) = \sum_{i=1}^{20} \alpha_i \frac{P_{rel}(i,j)}{p_i}, \tag{1.4}$$

This leads to the following scoring function.

$$score(\alpha, j) = \log \frac{\sum_{i=1}^{20} \alpha_i P(i \rightarrow j)}{p_j} = \log \sum_{i=1}^{20} \alpha_i \frac{P_{rel}(i,j)}{p_i p_j} \tag{1.5}$$

Again, summing up Eq. (1.5) over all aligned positions yields a score for the whole alignment. Similar to sequence alignment, dynamic programming can be used to generate an optimal alignment between one sequence and one profile for a scoring function defined in Eq. (1.3) or Eq. (1.5).

1.4.5 Scoring Function for Profile-Profile Alignment and Comparison

The sequence-profile scoring function defined in Eq. (1.5) can be extended to score a profile-profile alignment. Let X and Y be two aligned profile columns with amino acid probability distribution α_i ($i = 1, 2, \ldots, 20$) and β_j ($j = 1, 2, \ldots, 20$), respectively. The following log average score, which is a generalization of Eq. (1.5), can be used to estimate the similarity between these two profile columns.

$$LogAverageSco(\alpha, \beta) = \log \sum_{i=1}^{20} \sum_{j=1}^{20} \alpha_i \beta_i \frac{P_{rel}(i,j)}{p_i p_j} \tag{1.6}$$

Some methods also use the following average mutation score to measure the similarity of these two profile columns.

$$AverageSco(\alpha, \beta) = \sum_{i=1}^{20} \sum_{j=1}^{20} \alpha_i \beta_i \log \left(\frac{P_{rel}(i,j)}{p_i p_j} \right) \tag{1.7}$$

Besides the scoring functions defined in Eqs. (1.6) and (1.7), the following dot product and Jensen-Shannon scores are also proposed in literature [56].

Dot product Calculating dot product is the simplest and fastest approach to compare two profile columns [57]. This method calculates the similarity of two aligned profile columns as follows.

$$DotProductSco(\alpha, \beta) = \sum_{i=1}^{20} \alpha_i \beta_i \tag{1.8}$$

This can be interpreted as the probability of identical amino acids being produced from distributions α and β independently. A variant of this scoring function is to calculate dot product using log-odds values as follows.

$$DotOddSco(\alpha, \beta) = \sum_{i=1}^{20} \log(\alpha_i/p_i)\log(\beta_i/p_i) \tag{1.9}$$

Jensen-Shannon function This scoring function was introduced by Yona and Levitt [56], which measures similarity of two probability distributions using information theory. The similarity measure is based only on the observed probability distributions, so it is independent of any evolutionary models. The similarity score of two profile columns is defined as a combination of their statistical similarity and the significance of the statistical similarity. In particular, the scoring function involves the calculation of a divergence score,

$$DivSco(\alpha, \beta) = \frac{1}{2}\left[\sum_{i=1}^{20} \alpha_i \log\left(\frac{\alpha_i}{(\alpha_i + \beta_i)/2}\right) + \sum_{i=1}^{20} \beta_i \log\left(\frac{\beta_i}{(\alpha_i + \beta_i)/2}\right)\right] \tag{1.10}$$

and a significance score.

$$SigSco(\alpha, \beta) = \frac{1}{2}\left[\sum_{i=1}^{20} \alpha_i \log\left(\frac{\alpha_i}{p_i}\right) + \sum_{i=1}^{20} \beta_i \log\left(\frac{\beta_i}{p_i}\right)\right] \tag{1.11}$$

For more profile-profile scoring functions and their comparison, please refer to [58]. Experimental results indicate that the log-odds-based scoring functions, such as DotOddSco and Jensen-Shannon, more likely to perform better than many others [56].

1.5 Contribution of This Book

To significantly advance remote homology detection and fold recognition, this book focuses on profile-profile alignment, although the method presented in this book can be easily adapted for sequence-profile alignment. In particular, this book describes a Markov Random Fields (MRFs) representation of sequence profile. That is, we use MRF to model a multiple sequence alignment (MSA) of close sequence homologs. Compared to Hidden Markov Model (HMM) that can only model local-range residue correlation, MRFs can model long-range residue interactions (e.g., residue co-evolution) and thus, encodes global information in a protein. In particular, MRF is a graphical model encoding a probability distribution over the MSA by a graph and a set of preset statistical functions. A node in the MRF corresponds to one column in the MSA and one edge specifies correlation between two columns. Each node is associated with a function describing position-specific amino acid mutation pattern. Similarly, each edge is associated with a function describing correlated mutation statistics between two columns. With profile MRF representation, alignment of two proteins or protein families becomes that of two MRFs. To align two MRFs, a scoring function or alignment potential is needed to measure the similarity of two MRFs. We use a scoring function consists of both node alignment potential and edge alignment potential, which measure the node (i.e., amino acid) similarity and edge (i.e., interaction pattern) similarity, respectively.

It is computationally challenging to optimize a scoring function containing edge alignment potential. To deal with this, we formulate MRF-MRF alignment as an integer programming problem and then develop an Alternative Direction Method of Multipliers (ADMM) [59] algorithm to solve it efficiently to a suboptimal solution. ADMM divides the MRF alignment problem into two tractable sub-problems and then iteratively solve them until they reach consistent solutions. Experiments show that this MRF-MRF alignment method, denoted as MRFalign, can generate more accurate alignments and is also much more sensitive than others in detecting remote homologs. MRFalign works particularly well on mainly-beta proteins.

The most relevant work is Cowen's MRFy/SMURF methods for fold recognition of beta proteins [60, 61]. Nevertheless, our MRFalign method is significantly different from MRFy/SMURF in a few aspects: (1) MRFy/SMURF builds an MRF based upon multiple structure alignment instead of multiple sequence alignment (MSA). As such, it cannot apply to sequence-based homology detection in the absence of native structures. In contrast, our method builds MRFs purely based upon MSA and thus, applies to sequence-based protein alignment and homology detection; (2) MRFy/SMURF can only align a single primary sequence to an MRF, while our method aligns two MRFs to yield higher sensitivity; and (3) MRFy/SMURF does not take into consideration residue co-evolution information. This difference requires us to develop totally new methods to build MRFs from MSA, measure similarity of two MRFs, and optimize the MRF-MRF alignment potential.

References

1. Brent, M.R.: Steady progress and recent breakthroughs in the accuracy of automated genome annotation. Nat. Rev. Genet. **9**(1), 62–73 (2008)
2. Consortium, G.O.: The gene ontology project in 2008. Nucleic Acids Res. **36**(suppl 1), D440–D444 (2008)
3. Watson, J.D., Laskowski, R.A., Thornton, J.M.: Predicting protein function from sequence and structural data. Curr. Opin. Struct. Biol. **15**(3), 275–284 (2005)
4. Ginalski, K.: Comparative modeling for protein structure prediction. Curr. Opin. Struct. Biol. **16**(2), 172–177 (2006)

5. Flöckner, H., et al.: Progress in fold recognition. Proteins Struct. Funct. Bioinf. **23**(3), 376–386 (1995)
6. Eddy, S.R.: Profile hidden Markov models. Bioinformatics **14**(9), 755–763 (1998)
7. Baker, D., Sali, A.: Protein structure prediction and structural genomics. Science **294**(5540), 93–96 (2001)
8. Šali, A., et al.: Evaluation of comparative protein modeling by MODELLER. Proteins Struct. Funct. Bioinf. **23**(3), 318–326 (1995)
9. Fariselli, P., et al.: The WWWH of remote homolog detection: the state of the art. Briefings Bioinf. **8**(2), 78–87 (2007)
10. Wan, X.-F., Xu, D.: Computational methods for remote homolog identification. Curr. Protein Pept. Sci. **6**(6), 527–546 (2005)
11. Madera, M., Gough, J.: A comparison of profile hidden Markov model procedures for remote homology detection. Nucleic Acids Res. **30**(19), 4321–4328 (2002)
12. Jones, D.T., Taylor, W.R., Thornton, J.M.: The rapid generation of mutation data matrices from protein sequences. Comput. Appl. Biosci. CABIOS **8**(3), 275–282 (1992)
13. Grigoriev, I.V., Kim, S.-H.: Detection of protein fold similarity based on correlation of amino acid properties. Proc. Natl. Acad. Sci. **96**(25), 14318–14323 (1999)
14. Deschavanne, P., Tuffery, P.: Exploring an alignment free approach for protein classification and structural class prediction. Biochimie **90**(4), 615–625 (2008)
15. Jaakkola, T., Diekhans, M., Haussler, D.: A discriminative framework for detecting remote protein homologies. J. Comput. Biol. **7**(1–2), 95–114 (2000)
16. Kuang, R., et al.: Profile-based string kernels for remote homology detection and motif extraction. J. Bioinf. Comput. Biol. **3**(03), 527–550 (2005)
17. Leslie, C.S., Eskin, E., Noble, W.S.: The spectrum kernel: a string kernel for SVM protein classification. In: Pacific Symposium on Biocomputing (2002)
18. Liao, L., Noble, W.S.: Combining pairwise sequence similarity and support vector machines for detecting remote protein evolutionary and structural relationships. J. Comput. Biol. **10**(6), 857–868 (2003)
19. Jaakkola, T., Diekhans, M., Haussler, D.: Using the Fisher kernel method to detect remote protein homologies. In: ISMB (1999)
20. Leslie, C.S., et al.: Mismatch string kernels for discriminative protein classification. Bioinformatics **20**(4), 467–476 (2004)
21. Byvatov, E., Schneider, G.: Support vector machine applications in bioinformatics. Appl. Bioinf. **2**(2), 67–77 (2002)
22. Jebara, T.: Machine Learning: Discriminative and Generative. Springer, Berlin (2004)
23. Balakrishnan, S., et al.: Learning generative models for protein fold families. Proteins Struct. Funct. Bioinf. **79**(4), 1061–1078 (2011)
24. Thomas, J., Ramakrishnan, N., Bailey-Kellogg, C.: Protein design by sampling an undirected graphical model of residue constraints. IEEE/ACM Trans. Comput. Biol. Bioinf. **6**(3), 506–516 (2009)
25. Shen, H.-B., Chou, K.-C.: Ensemble classifier for protein fold pattern recognition. Bioinformatics **22**(14), 1717–1722 (2006)
26. Tan, A., Gilbert, D., Deville, Y.: Multi-class protein fold classification using a new ensemble machine learning approach (2003)
27. Dehzangi, A., Phon-Amnuaisuk, S., Dehzangi, O.: Using random forest for protein fold prediction problem: an empirical study. J. Inf. Sci. Eng. **26**(6), 1941–1956 (2010)
28. Lundström, J., et al.: Pcons: a neural-network-based consensus predictor that improves fold recognition. Protein Sci. **10**(11), 2354–2362 (2001)
29. McGuffin, L.J., Jones, D.T.: Improvement of the GenTHREADER method for genomic fold recognition. Bioinformatics **19**(7), 874–881 (2003)
30. Zakeri, P., et al.: Protein fold recognition using geometric kernel data fusion. Bioinformatics btu118 (2014)

31. Do, C.B., Gross, S.S., Batzoglou, S.: CONTRAlign: discriminative training for protein sequence alignment. In: Research in Computational Molecular Biology. Springer, Berlin (2006)
32. Ding, C.H., Dubchak, I.: Multi-class protein fold recognition using support vector machines and neural networks. Bioinformatics **17**(4), 349–358 (2001)
33. Dong, Q., Zhou, S., Guan, J.: A new taxonomy-based protein fold recognition approach based on autocross-covariance transformation. Bioinformatics **25**(20), 2655–2662 (2009)
34. Sharma, A., et al.: A feature extraction technique using bi-gram probabilities of position specific scoring matrix for protein fold recognition. J. Theor. Biol. **320**, 41–46 (2013)
35. Smith, T.F., Waterman, M.S.: Comparison of biosequences. Adv. Appl. Math. **2**(4), 482–489 (1981)
36. Pearson, W.R.: Searching protein sequence libraries: comparison of the sensitivity and selectivity of the Smith-Waterman and FASTA algorithms. Genomics **11**(3), 635–650 (1991)
37. Altschul, S.F., et al.: Gapped BLAST and PSI-BLAST: a new generation of protein database search programs. Nucleic Acids Res. **25**(17), 3389–3402 (1997)
38. Pearson, W.R.: [5] Rapid and sensitive sequence comparison with FASTP and FASTA. Methods Enzymol. **183**, 63–98 (1990)
39. Henikoff, S., Henikoff, J.G.: Amino acid substitution matrices from protein blocks. Proc. Natl. Acad. Sci. **89**(22), 10915–10919 (1992)
40. Eddy, S.R.: HMMER: profile hidden Markov models for biological sequence analysis (2001)
41. Hughey, R., Krogh, A.: Hidden Markov models for sequence analysis: extension and analysis of the basic method. Comput. Appl. Biosci. CABIOS **12**(2), 95–107 (1996)
42. Morgenstern, B., et al.: DIALIGN: finding local similarities by multiple sequence alignment. Bioinformatics **14**(3), 290–294 (1998)
43. Probst, W.C., et al.: Sequence alignment of the G-protein coupled receptor superfamily. DNA Cell Biol. **11**(1), 1–20 (1992)
44. Söding, J.: Protein homology detection by HMM–HMM comparison. Bioinformatics **21**(7), 951–960 (2005)
45. Tomii, K., Akiyama, Y.: FORTE: a profile–profile comparison tool for protein fold recognition. Bioinformatics **20**(4), 594–595 (2004)
46. Heger, A., Holm, L.: Picasso: generating a covering set of protein family profiles. Bioinformatics **17**(3), 272–279 (2001)
47. Moult, J.: A decade of CASP: progress, bottlenecks and prognosis in protein structure prediction. Curr. Opin. Struct. Biol. **15**(3), 285–289 (2005)
48. Pruitt, K.D., Tatusova, T., Maglott, D.R.: NCBI reference sequence (RefSeq): a curated non-redundant sequence database of genomes, transcripts and proteins. Nucleic Acids Res. **33** (suppl 1), D501–D504 (2005)
49. Bates, P.A., et al.: Enhancement of protein modeling by human intervention in applying the automatic programs 3D-JIGSAW and 3D-PSSM. Proteins Struct. Funct. Bioinf. **45**(S5), 39–46 (2001)
50. Koonin, E.V., Wolf, Y.I., Aravind, L.: Protein fold recognition using sequence profiles and its application in structural genomics. Adv. Protein Chem. **54**, 245–275 (2000)
51. Eddy, S.R.: Hidden markov models. Curr. Opin. Struct. Biol. **6**(3), 361–365 (1996)
52. Bateman, A., et al.: The Pfam protein families database. Nucleic Acids Res. **32**(suppl 1), D138–D141 (2004)
53. Bateman, A., et al.: The Pfam protein families database. Nucleic Acids Res. **30**(1), 276–280 (2002)
54. Gough, J., Chothia, C.: SUPERFAMILY: HMMs representing all proteins of known structure. SCOP sequence searches, alignments and genome assignments. Nucleic Acids Res. **30**(1), 268–272 (2002)
55. Ma, J., et al.: MRFalign: protein homology detection through alignment of Markov random fields. PLoS Comput. Biol. **10**(3), e1003500 (2014)
56. Yona, G., Levitt, M.: Within the twilight zone: a sensitive profile-profile comparison tool based on information theory. J. Mol. Biol. **315**(5), 1257–1275 (2002)

57. Rychlewski, L., Zhang, B., Godzik, A.: Fold and function predictions for fold and function predictions for. Fold Des. **3**(4), 229–238 (1998)
58. Wang, G., Dunbrack, R.L.: Scoring profile-to-profile sequence alignments. Protein Sci. **13**(6), 1612–1626 (2004)
59. Boyd, S., et al.: Distributed optimization and statistical learning via the alternating direction method of multipliers. Found. Trends® Mach. Learn. **3**(1), 1–122 (2011)
60. Daniels, N.M., et al.: SMURFLite: combining simplified Markov random fields with simulated evolution improves remote homology detection for beta-structural proteins into the twilight zone. Bioinformatics **28**(9), 1216–1222 (2012)
61. Daniels, N.M., et al.: MRFy: remote homology detection for beta-structural proteins using Markov random fields and stochastic search. In: Proceedings of the International Conference on Bioinformatics, Computational Biology and Biomedical Informatics. ACM (2013)

Chapter 2
Method

Abstract This chapter describes a novel MRF-based method for homology detection and fold recognition. In particular, it covers how to build an MRF model for a protein sequence, how to score the similarity of two MRF models and the similarity between one MRF model and one native structure, and finally an alternating direction method of multipliers (ADMM) method that can optimize the scoring function.

Keywords Markov random fields (MRF) · Hidden Markov models (HMM) · Alternating direction method of multipliers (ADMM) · Mutual information · Residue co-evolution

2.1 Modeling a Protein Family Using Markov Random Fields

Given a protein sequence, we run PSI-BLAST [1] with 5 iterations and E-value cutoff 0.001 to find its sequence homologs and then build their multiple sequence alignment (MSA). We can use a multivariate random variable $X = (X_1, X_2, \ldots, X_N)$, where N is the number of columns (or the MSA length), to model the MSA. Here each X_i is a finite discrete random variable representing the amino acid at column i in the MSA, taking values from 1 to 21, corresponding to 20 amino acids and gap. The occurring probability of the whole MSA can be modeled by an Markov Random Field (MRF), which is a function of X. MRF is an undirected graph that can be used to model a set of correlated random variables. As shown in Fig. 2.1, an MRF node represents one column in the MSA and an edge represents the correlation between two columns. Here we ignore very short-range residue correlation since it is not very informative. An MRF consists of two types of functions: $\phi(X_i)$ and $\psi(X_i, X_k)$, where $\phi(X_i)$ is an amino acid preference function for node i and $\psi(X_i, X_k)$ is a pairwise amino acid preference function for edge (i, k)

© The Author(s) 2015
J. Xu et al., *Protein Homology Detection Through Alignment of Markov Random Fields*, SpringerBriefs in Computer Science, DOI 10.1007/978-3-319-14914-1_2

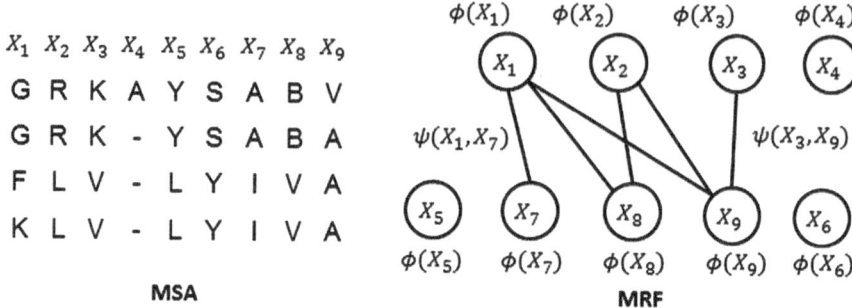

Fig. 2.1 Modeling a multiple sequence alignment (*left*) by Markov random fields (*right*)

that reflects interaction between two nodes. Then, the probability of observing a particular protein sequence X can be calculated as follows.

$$P(X) = \frac{1}{Z} \Pi_i \phi(X_i) \Pi_{(i,k)} \psi(X_i, X_k), \qquad (2.1)$$

where Z is the normalization factor (i.e., partition function). If $P(X)$ is Gaussian, this MRF is called Gaussian Graphical Model (GGM). In this case, X_i shall be interpreted as a 21-dimension binary vector, in which each element corresponds to one amino acid type or gap.

We use two kinds of information in MRFs for their alignment. One is the occurring probability of 20 amino acids and gap at each node (i.e., each column in MSA), which can also be interpreted as the marginal probability at each node. The other is the correlation between two nodes, which can be interpreted as interaction strength of two MSA columns.

2.2 Estimating the Parameters of Markov Random Fields

In this section, we describe how to estimate the parameters in the ψ function of the MRF. As stated above, ψ is a pairwise amino acid preference function for edge (i,k) reflecting interactions between two nodes (residues). A couple of strategies can be used to estimate residue interaction strength. One is to estimate residue co-evolution strength from the MSA and then use this co-evolution strength as residue interaction strength. Mutual information (MI) is a simple and popular way to estimate residue co-evolution strength. However, mutual information (MI) is only a local statistics (i.e., the MI of 2 positions is calculated independent of the other positions), so MI is not very accurate in estimating residue co-evolution strength. Instead of using MI, we can use direct information (DI), which is a global statistics (i.e., measuring the residue co-evolution strength of two positions considering other positions). DI can be calculated by some contact prediction programs such PSICOV, Evfold, plmDCA [2–4] as residue co-evolution. PSICOV assumes that $P(X)$ is a Gaussian distribution function and calculates the correlation between two columns by inverse covariance matrix. By contrast, plmDCA does not assume a Gaussian distribution and is more efficient and also slightly more accurate. Generally speaking, these programs are time-consuming.

The reliability of mutual information (MI) or direct information (DI) [2] depends on the number of non-redundant sequence homologs. When there are few sequence homologs, the resulting MI or DI is not very accurate. Therefore, it is not enough to only use residue co-evolution strength to estimate residue interaction strength. We can use other contact prediction programs such as PhyCMAP [4] which integrates both residue col-evolution information, PSI-BLAST sequence profile and others to predict the probability of two residues in contact. PhyCMAP works much better than PSICOV and Evfold when proteins under study have a small number of sequence homologs [4].

In this work, we use predicted inter-residue Euclidean distance to reflect interaction strength of two residues. This is based upon an assumption that two spatially-close residues tend to have strong interaction. We predict the inter-residue distance using sequence information such as mutual information and its power series, PSI-BLAST sequence profile and other protein features. See [5] for more details. Below we briefly describe how to predict inter-residue distance from sequence information using probabilistic neural networks (PNN).

We discretize $C_\alpha - C_\alpha$ distance into 13 bins (3–4, 4–5, 5–6, ..., 14–15, and >15 Å). Each bin is also called a label. Given a protein and a pair of two residues i and j, let d_k denote the bin into which their distance falls, and x_k denote the protein feature vector consisting of some position-specific sequence profile information and also mutual information between two positions. We would like to estimate the probability of observing d_k given the feature vector x_k. That is, instead of only considering the most possible distance labels assigned to each pair of nodes (residues), we would like to estimate the probability distribution of d_k. The reason is that the predicted distance probability distribution is more informative than a single predicted value.

Formally, let $p_\theta(d_k|x_k)$ be the probability of the distance label d_k conditioned on the feature vector x_k Meanwhile, θ is the model parameter vector. We estimate $p_\theta(d_k|x_k)$ as follows:

$$p_\theta(d_k|x_k) = \frac{\exp(L_\theta(d_k, x_k))}{Z_\theta(x_k)} \tag{2.2}$$

where $Z_\theta(x_k) = \sum_d \exp(L_\theta(d, x_k))$, is the partition function and $L_\theta(d, x_k)$ is a two-layer neural network. Figure 2.2 shows an example of the neural network with three and five neurons in the first and second hidden layers, respectively. Each neuron is a sigmoid function. The function $L_\theta(d_k, x_k)$ can be calculated as,

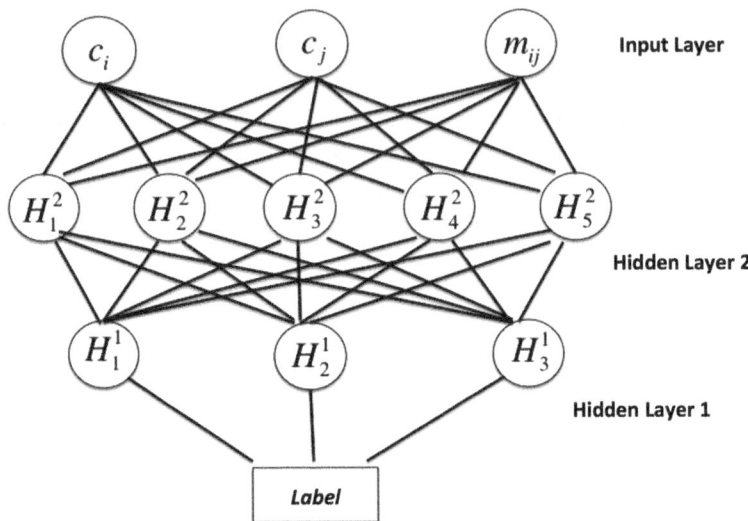

Fig. 2.2 Illustration of probabilistic neural networks (PNN), in which c_i and c_j are the protein sequence features centered at the ith and jth residues, respectively, and m_{ij} (mutual information) is the feature that must be fetched at the ith and jth residues simultaneously

$$L_\theta(d_k, x_k) = \sum_{g_1=1}^{G_1} \theta^0_{d_k, g_1} h(\sum_{g_1, g_2}^{G_2} \theta^1_{g_1, g_2} h(<\theta^2_{g_2}, x_k >)) \qquad (2.3)$$

where G_1 and G_2 are the number of gates in the two hidden layers, $<\cdot, \cdot>$ denotes the inner product of two vectors, $\theta^2_{g_2}$ is the weight factor of the g_2th neuron in the second layer; $\theta^1_{g_1, g_2}$ is the weight connecting the first layer and the second layer. $\theta^0_{d_k, g_1}$ is the weight connecting the first layer and the labels.

In current implementation, our neural network contains two hidden layers. The first hidden layer (i.e., the layer connecting to the input layer) contains 100 neurons, and the second hidden layer (i.e., the layer connecting to the output layer) has 40 neurons. This neural network is similar to what is used by the Zhou group [6] for inter-residue contact prediction, which uses 100 and 30 neurons in the two hidden layers, respectively. The Zhou group has shown that using two hidden layers can obtain slightly better performance than using a single hidden layer. The input layer of our network has about 600 features, so in total, our neural network has between 60,000 and 70,000 parameters to be trained. We use the maximum likelihood method to train the model parameter θ and to determine the number of neurons in each hidden layer by maximizing the occurring probability of native $C_\alpha - C_\alpha$ distance in a set of training proteins. It is challenging to maximize the objective function Eq. (2.2) since it is non-convex and a large amount of training data is used. We use a limited-memory BFGS method [7] to fulfill this. We generated an initial solution randomly and then ran the training algorithm on a supercomputer for about a couple of weeks. Our training algorithm terminated when the probability of either the training set or the validation set did not improve any more. Note that all the model parameters are learned from the training set but not the validation set. The validation set, combined with the training set, is only used to determine when our training algorithm shall terminate. Our training algorithm usually terminates after 3,000 iterations. We also reran our training algorithm starting from nine initial solutions and did not observe explicit performance difference among these runs. See our work on EPAD [5] for more details.

We use two kinds of input features in this neural network model: PSI-BLAST sequence profile and residue co-evolution. One is context-specific sequence profile for a small sequence segment centered at one specific residue in question. The sequence profile is generated by running PSI-BLAST on the NR database with 5 iterations and an E-value of 0.001. The other feature we used is residue co-evolution information. Mutual information is a classical method to measure residue co-evolution strength. However, mutual information cannot differentiate direct from indirect interactions. For example, when residue a has strong interaction with b and b has strong interaction with residue c, it is likely that residue a also has interaction with c. In order to reduce the impact of this kind of indirect information, some global statistical methods such as Graphical Lasso [3] and Pseudo-likelihood [8, 9] methods are proposed to estimate residue co-evolution strength. However, these methods are time-consuming. In this work, to account for chaining effect of residue coupling, we use the powers of the mutual information matrix. In particular, let MI denotes the mutual information matrix, we use MI^k where k ranges from 2 to 11 to estimate the chaining effect.

2.3 Scoring Similarity of Two Markov Random Fields

This section will introduce how to align two proteins by aligning their corresponding MRFs. As shown in the left picture of Fig. 2.3, building an alignment is equivalent to finding a unique path from the left-top corner to the right-bottom corner. For each vertex along the path, we need a score to measure how good it is to transit to the next vertex. That is, we need to measure how similar two nodes of the two MRFs are. We call this kind of scoring function node alignment potential. Second, in addition to measure the similarity two aligned MRF nodes, we want to quantify the similarity between two MRF edges. For example, in the right picture of Fig. 2.3 residues "L" and "S" of the first protein are aligned to residues "A" and "Q" of the 2nd protein, respectively. We would like to estimate how good it is to align the pair (L, S) to the pair (A, Q). This pairwise similarity function is a function of two MRF edges and we call it edge alignment potential. When the edge alignment potential is used to score the similarity of two MRFs, Viterbi algorithm or simple dynamic programming cannot be used to find the optimal alignment. It can

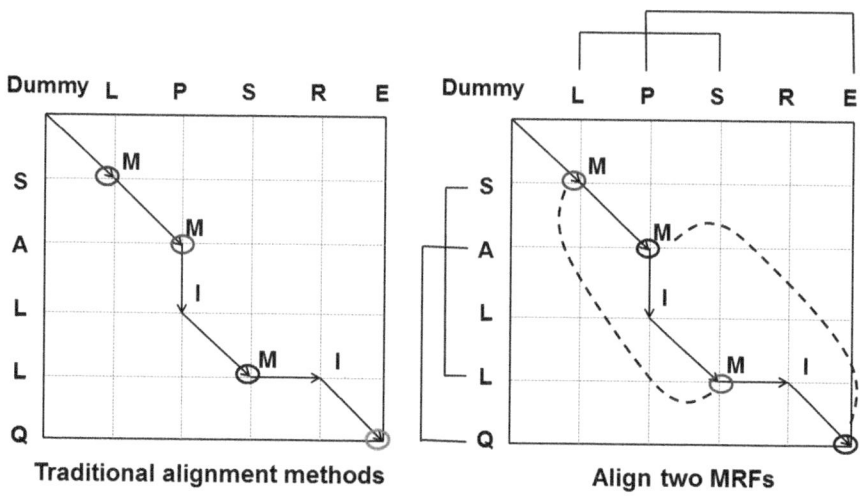

Fig. 2.3 Traditional alignment methods (*left*) and our MRFalign method (*right*)

be proved that when edge alignment potential is considered and gaps are allowed the MRF-MRF alignment problem is NP hard [10]. In this work, we will describe an ADMM [11] algorithm to quickly find a suboptimal alignment of two MRFs. Although suboptimal, empirically the resultant alignment has very good accuracy.

2.4 Node Alignment Potential of Markov Random Fields

Given an alignment path, its node alignment potential is the accumulative potential of all the vertices along the path. In particular, the node alignment potential is a function of two MRF nodes and measures the similarity of two aligned positions using local protein features. We use a Conditional Neural Fields (CNF) [12] method to develop such a node alignment potential, using a procedure very similar to what is described in the protein threading paper [13, 14]. Briefly speaking, we estimate the occurring probability of an alignment A between two proteins T and S as follows.

$$P(A|T,S) = e^{\sum_{(i,j,u)\in A} E_u(T_i,S_j)}/Z(T,S) \qquad (2.4)$$

where $Z(T,S)$ is a normalization factor summarizing all the possible alignments between T and S, and $E_u(T_i,S_j)$ is a neural network with one hidden layer that calculates the log-likelihood of a vertex (i,j,u) in the alignment path, where i is a node in T, j a node in S, and u a state. When u is a match state, E_u takes as input the sequence profile context of two nodes i and j, denoted as T_i and S_j, respectively, and yields the log-likelihood of these two nodes being matched. When u is an insertion state, it takes as input at the sequence profile context of one node and yields the log-likelihood of this node being an insertion. The sequence profile context of node i is a $21 \times (2w+1)$ matrix where $w = 5$, consisting of the marginal probability of 20 amino acids and gap at $2w+1$ nodes indexed by $i-w$, $i-w+1, \ldots, i, i+1, \ldots, i+w$. In case that one column does not exist (when $i \leq w$ or $i+w > N$), zero is used. See Fig. 2.4 for an illustration of how to use neural networks to estimate local similarity of two MRF nodes.

We train the parameters in E_u by maximizing the occurring probability of a set of reference alignments, which are generated by a structure alignment tool DeepAlign [15]. That is, we optimize the model parameters so that the structure alignment of one training protein pair has the largest probability among all possible alignments. Notice that by using neural networks the objective function of Eq. (2.4) is neither concave nor log-concave, so it is challenging to find globally optimal solution. Here we use the Limited memory BFGS (L-BFGS) algorithm to solve it to suboptimal. To obtain a good solution, we run L-BFGS several times starting from different initial solutions and return the best suboptimal solution. A L_2-norm regularization factor, which is determined by fivefold cross validation, is used to restrict the search space of model parameters to avoid over-fitting.

Let $\theta_{i,j}^u$ denote the local alignment potential of a vertex (i, j, u) in the alignment path. We calculate $\theta_{i,j}^u$ from E_u as follows.

$$\theta_{i,j}^u = E_u(T_i, S_j) - \text{Exp}(E_u) \tag{2.5}$$

where $\text{Exp}(E_u)$ is the expected value of E_u, which depends only on the alignment state but not any specific protein pair. It is used to offset the effect of the background, which is the log-likelihood yielded by E_u for any randomly chosen node

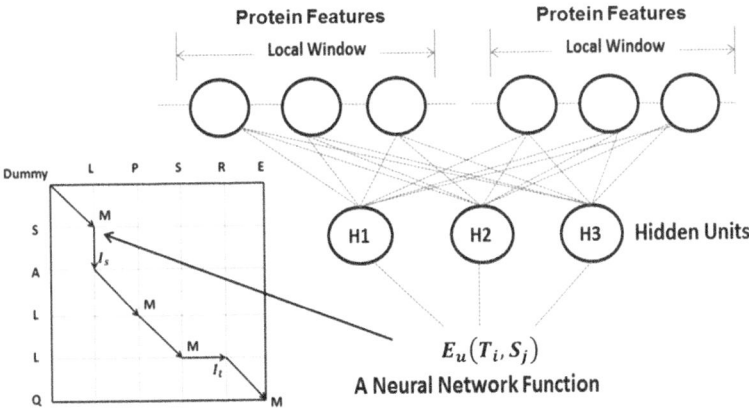

Fig. 2.4 An example neural network for calculating node alignment potential, in which there is one hidden layer. The function takes features from both proteins as input and yields one log-likelihood score

pairs (or nodes). We can calculate the reference alignment probability E_u in Eq. (2.5) by randomly sampling a set of protein pairs, each with the same lengths as the sequence S and template T, respectively, and then estimating the probability of alignment A based upon these randomly sampled protein pairs. As long as we generate sufficient number of samplings, we shall be able to approximate $\text{Exp}(E_u)$ very well.

To perform protein homology detection, only sequence and profile based features are included. For the sequence similarity features, we use BLOSUM 80 and BLOSUM 62 matrix [16], structure-derived amino acid substitution matrix [17], and statistical potential derived amino acid similarity matrix [18]. We also use sequence-profile similarity score and profile-profile similarity score as described in section Introduction. We do not use an affine gap penalty, but a context-specific gap penalty which is parameterized by amino acid type, predicted secondary structure type and position-specific profile information.

2.5 Edge Alignment Potential of Markov Random Fields

The edge alignment potential calculates the similarity of two edges, one from each MRF, based upon the interaction strength of two ends in one edge. As mentioned above, we use predicted distance probability distribution based on the features of two nodes to estimate their interaction strength. Let d_{ik}^T be a random variable for the Euclidean distance between two residues at i and k and d_{jl}^S is defined similarly. Let $\theta_{i,k,j,l}$ denote the alignment potential between edge (i, k) in T and edge (j, l) in S. As shown in Fig. 2.5 we calculate $\theta_{i,k,j,l}$ as follows.

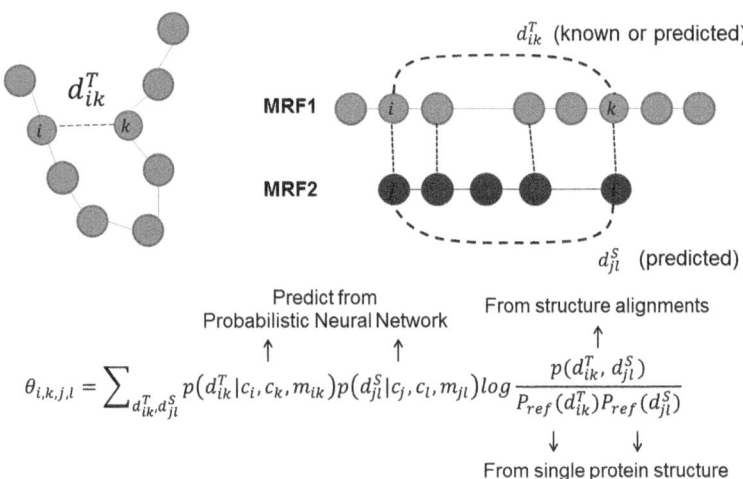

$$\theta_{i,k,j,l} = \sum_{d_{ik}^T, d_{jl}^S} p(d_{ik}^T | c_i, c_k, m_{ik}) p(d_{jl}^S | c_j, c_l, m_{jl}) log \frac{p(d_{ik}^T, d_{jl}^S)}{P_{ref}(d_{ik}^T) P_{ref}(d_{jl}^S)}$$

Fig. 2.5 Illustration of edge alignment potential for MRF-MRF alignment

$$\theta_{i,k,j,l} = \sum_{d_{ik}^T, d_{jl}^S} p\left(d_{ik}^T | c_i, c_k, m_{ik}\right) p\left(d_{jl}^S | c_j, c_l, m_{jl}\right) \log \frac{p(d_{ik}^T, d_{jl}^S)}{P_{ref}(d_{ik}^T) P_{ref}(d_{jl}^S)} \tag{2.6}$$

where $p\left(d_{ik}^T | c_i, c_k, m_{ik}\right)$ is the probability of two nodes i and k in T interacting at distance d_{ik}^T; $p\left(d_{jl}^S | c_j, c_l, m_{jl}\right)$ is the probability of two nodes j and l in S interacting at distance d_{jl}^S; $p\left(d_{ik}^T, d_{jl}^S\right)$ is the probability of one distance d_{ik}^T being aligned to another distance d_{jl}^S in reference alignments; and $P_{ref}\left(d_{ik}^T\right)$ $\left(P_{ref}\left(d_{jl}^S\right)\right)$ is the background probability of observing d_{ik}^T (d_{jl}^S) in a protein structure. Meanwhile x_i and x_k are position-specific features centered at the ith and kth residues, respectively, and m_{ik} represents the mutual information between the ith and kth columns in the multiple sequence alignment.

Compared to contact-based potentials, here we use interaction at a given distance to obtain a higher-resolution description of the residue interaction pattern, as shown in Fig. 2.5. Therefore, this edge alignment potential is more informative and thus, may lead to better alignment accuracy and homology detection rate.

Now we explain how to calculate each term in Eq. (2.6). $P_{ref}\left(d_{ik}^T\right)$ $\left(P_{ref}\left(d_{jl}^S\right)\right)$ can be calculated by simple counting on a set of non-redundant protein structures, e.g., PDB25. Similar to $P_{ref}\left(d_{ik}^T\right)$, $P\left(d_{ik}^T, d_{jl}^S\right)$ can also be calculated by simple counting on a set of non-redundant reference alignments. That is, we randomly choose a set of protein pairs such that two proteins in each pair are similar at least at the fold level. Then we generate their reference alignment (i.e., structure alignments) using a structure alignment tool DeepAlign [15] and finally do simple counting to estimate $p\left(d_{ik}^T, d_{jl}^S\right)$. In order to use simple counting, we discretize inter-residue distance into 12 intervals: <4, 4–5, 5–6, ..., 14–15, and >15 Å.

As explained in the previous section, we predict $p\left(d_{ik}^T | c_i, c_k, m_{ik}\right)$ using a probabilistic neural network (PNN) implemented in our context-specific distance-dependent statistical potential package EPAD. EPAD takes as input sequence profile contexts and mutual information and then yields inter-residue distance probability distribution. See the EPAD paper [5] for the technical details. The EPAD package has been blindly tested in CASP10 for template free modeling. The CASP10 results show that EPAD can successfully fold some targets with unusual fold (according to the CASP10 Free Modeling assessor Dr. BK Lee). Our large-scale experimental test also indicates EPAD is much better than those context-independent distance-based pairwise potentials such as DOPE [19], RW [20] and DFIRE [21] in ranking protein decoys [5].

2.6 Scoring Similarity of One Markov Random Fields and One Template

This MRF-based alignment method can also be applied to protein threading. In this scenario, one of the two proteins under alignment has solved 3D structure. Of course we can just simply use the node and edge alignment potentials described in previous sections to align one MRF to one solved structure. In order to use the native structure information in the protein with solved 3D structure, we may revise the alignment potentials as follows.

1. Instead of using predicted secondary structure and solvent accessibility, we may use their native information for the protein with solved 3D structure, which can be generated by DSSP [22].
2. Let T denote the protein with solved 3D structure. We can directly calculate the inter-residue distance for any residue pairs in T. That is, $p\left(d_{ik}^T|c_i, c_k, m_{ik}\right)$ reduces to a simple distribution that has probability 1 for the native distance between residues i and k and 0 otherwise. So, the edge alignment potential can be simplified as follows.

$$\theta_{i,k,j,l} = P\left(d_{ij}^T|c_i, c_j\right) = \sum_{d_{ij}^S} P(d_{ij}^T|d_{ij}^S)P(d_{ij}^S|c_i, c_j) \tag{2.7}$$

where d_{ij}^S represents the distance of the two sequence residues at the two aligned positions, $P\left(d_{ij}^T|d_{ij}^S\right)$ is the conditional probability of d_{ij}^T on d_{ij}^S and $P\left(d_{ij}^S|c_i, c_j\right)$ is the conditional probability of d_{ij}^S estimated from the contexts (denoted x_i and x_j) of the two sequence residues. In Eq. (2.7), $P\left(d_{ij}^T|d_{ij}^S\right) = \frac{P(d_{ij}^T, d_{ij}^S)}{P(d_{ij}^S)}$ where $P\left(d_{ij}^S\right)$ $\left(= P_{ref}\left(d_{ij}^T\right)\right)$ is the background probability, and $P\left(d_{ij}^T, d_{ij}^S\right)$ is the joint probability of the pairwise distances of two aligned residue pairs and can be calculated by simple statistics using a set of non-redundant protein structure alignments generated by a structure alignment tool such as DeepAlign.

2.7 Algorithms for Aligning Two Markov Random Fields

As mentioned before, an alignment can be represented as a path in the alignment matrix, which encodes an exponential number of paths. We can use a set of $3N_1N_2$ binary variables $z_{i,j}^u$ to define a path, where N_1 and N_2 are the lengths of the two MSAs, (i, j) is an entry in the alignment matrix and u the associated state. Meanwhile, $z_{i,j}^u$ is equal to 1 if the alignment path passes (i, j) with state u. Therefore, the problem of finding the best alignment between two MRFs can be formulated as the following quadratic optimization problem.

$$\text{(P1)} \quad \max_z \sum_{i,j,u} \theta_{i,j}^u z_{i,j}^u + \frac{1}{L} \sum_{i,j,k,l,u,v} \theta_{i,j,k,l}^{uv} z_{i,j}^u z_{k,l}^v \tag{2.8}$$

where $\theta_{i,j}^u$ and $\theta_{i,j,k,l}^{uv}$ are node and edge alignment potentials as described in previous section. Meanwhile, $\theta_{i,j,k,l}^{uv}$ is equal to 0 if either u or v is not a match state. L is the alignment length and $1/L$ is used to make the accumulative node and edge potential have similar scale. Note that L is unknown and we will describe how to determine it later in this section. Finally, the solution of P1 is also subject to the constraint that all those $z_{i,j}^u$ with value 1 shall form a valid alignment path. This constraint shall be enforced to all the optimization problems described in this section.

It is computationally intractable to find the optimal solution of P1. Below we present an Alternating Direction Method of Multipliers (ADMM) method that can efficiently solve this problem to suboptimal. See [11] for a tutorial of the ADMM method. To use ADMM, we rewrite P1 as follows by making a copy of Z to y, but without changing the solution space.

$$\text{(P2)} \quad \max_{z,y} \sum_{i,j,u} \theta_{i,j}^u z_{i,j}^u + \frac{1}{L} \sum_{i,j,k,l,u,v} \theta_{i,j,k,l}^{uv} z_{i,j}^u y_{k,l}^v \tag{2.9}$$
$$s.t. \ \forall k,l,v, \quad z_{k,l}^v = y_{k,l}^v$$

Problem P2 can be augmented by adding a term to penalize the difference between z and y.

$$\text{(P3)} \quad \max_{z,y} \sum_{i,j,u} \theta_{i,j}^u z_{i,j}^u + \frac{1}{L} \sum_{i,j,k,l,u,v} \theta_{i,j,k,l}^{uv} z_{i,j}^u y_{k,l}^v - \frac{\rho}{2} \sum_{i,j,u} \left(z_{i,j}^u - y_{i,j}^u \right)^2 \tag{2.10}$$
$$s.t. \ \forall i,j,u, \quad z_{i,j}^u = y_{i,j}^u$$

P3 is equivalent to P2 and P1, but converges faster due to the penalty term. Here ρ is a hyper-parameter influencing the convergence rate of the algorithm. Empirically, setting ρ to a constant (=0.5) enables our algorithm to converge within 10 iterations for most protein pairs.

Adding the constraint $z_{i,j}^u = y_{i,j}^u$ using a Lagrange multiplier λ to Eq. (2.10), we have the following Lagrangian dual problem:

$$\text{(P4)} \quad \min_\lambda \max_{z,y} \sum_{i,j,u} \theta_{i,j}^u z_{i,j}^u + \frac{1}{L} \sum_{i,j,k,l,u,v} \theta_{i,j,k,l}^{uv} z_{i,j}^u y_{k,l}^v$$
$$+ \sum_{i,j,u} \lambda_{i,j}^u \left(z_{i,j}^u - y_{i,j}^u \right) - \frac{\rho}{2} \sum_{i,j,u} \left(z_{i,j}^u - y_{i,j}^u \right)^2 \tag{2.11}$$

It is easy to prove that P3 is upper bounded by P4. Now we will solve P4 and use its solution to approximate P3 and thus, P1.

Since both z and y are binary variables, the last term in Eq. (2.11) can be expanded as follows.

$$\frac{\rho}{2}\sum_{i,j,u}\left(z_{i,j}^{u} - y_{i,j}^{u}\right)^{2} = \frac{\rho}{2}\sum_{i,j,u}(z_{i,j}^{u} + y_{i,j}^{u} - 2z_{i,j}^{u}y_{i,j}^{u}) \tag{2.12}$$

For a fixed λ, we can split P4 into the following two sub-problems.

$$(\text{SP1}) \quad y^{*} = \text{argmax}\sum_{k,l,v}y_{k,l}^{v}C_{k,l}^{v} \tag{2.13}$$

where $C_{k,l}^{v} = \frac{1}{L}\sum_{i,j,u}\theta_{i,j,k,l}^{uv}z_{i,j}^{u} - \lambda_{k,l}^{v} - \frac{\rho}{2}\left(1 - 2z_{k,l}^{v}\right)$

$$(\text{SP2})\, z^{*} = \text{argmax}\sum_{i,j,u}z_{i,j}^{u}D_{i,j}^{u} \tag{2.14}$$

where $D_{i,j}^{u} = \theta_{i,j}^{u} + \sum_{k,l,v}\frac{1}{L}\theta_{i,j,k,l}^{uv}y_{k,l}^{v*} + \lambda_{i,j}^{u} - \frac{\rho}{2}(1 - y_{i,j}^{u*})$

The sub-problem SP1 optimizes the objective function with respect to y while fixing z, and the sub-problem SP2 optimizes the objective function with respect to z while fixing y. SP1 and SP2 do not contain any quadratic term, so they can be efficiently solved using the classical dynamic programming algorithm for sequence or HMM-HMM alignment.

In summary, we solve P4 using the following procedure

Step 1   Initialize z by aligning the two MRFs without the edge alignment potential, which can be done by dynamic programming. Accordingly, initialize L as the length of the initial alignment.

Step 2   Solve (SP1) first and then (SP2) using dynamic programming, each generating a feasible alignment.

Step 3   If the algorithm converges, i.e., the difference between z and y is very small or zero, stop here. Otherwise, we update the alignment length L as the length of the alignment just generated and the Lagrange multiplier λ using subgradient descent as in Eq. (2.15), and then go back to Step 2.

$$\lambda^{n+1} = \lambda^{n} - \rho(z^{*} - y^{*}) \tag{2.15}$$

Due to the quadratic penalty term in Eq. (2.10), this ADMM algorithm usually converges much faster and also yields better solutions than without this term. Empiric ally, it converges within 10 iterations for most protein pairs. See [11] for the convergence proof of a general ADMM algorithm.

References

1. Altschul, S.F., et al.: Gapped BLAST and PSI-BLAST: a new generation of protein database search programs. Nucleic Acids Res. **25**(17), 3389–3402 (1997)
2. Marks, D.S., et al.: Protein 3D structure computed from evolutionary sequence variation. PLoS ONE **6**(12), e28766 (2011)
3. Jones, D.T., et al.: PSICOV: precise structural contact prediction using sparse inverse covariance estimation on large multiple sequence alignments. Bioinformatics **28**(2), 184–190 (2012)
4. Wang, Z., Xu, J.: Predicting protein contact map using evolutionary and physical constraints by integer programming. Bioinformatics **29**(13), i266–i273 (2013)
5. Zhao, F., Xu, J.: A position-specific distance-dependent statistical potential for protein structure and functional study. Structure **20**(6), 1118–1126 (2012)
6. Faraggi, E., Xue, B., Zhou, Y.: Improving the prediction accuracy of residue solvent accessibility and real-value backbone torsion angles of proteins by guided-learning through a two-layer neural network. Proteins Struct. Funct. Bioinf. **74**(4), 847–856 (2009)
7. Malouf, R.: A comparison of algorithms for maximum entropy parameter estimation. In: Proceedings of the 6th Conference on Natural Language Learning, vol. 20. Association for Computational Linguistics (2002)
8. Ekeberg, M., Hartonen, T., Aurell, E.: Fast pseudo likelihood maximization for direct-coupling analysis of protein structure from many homologous amino-acid sequences. arXiv preprint arXiv:1401.4832 (2014)
9. Kamisetty, H., Ovchinnikov, S., Baker, D.: Assessing the utility of coevolution-based residue–residue contact predictions in a sequence-and structure-rich era. Proc. Natl. Acad. Sci. **110** (39), 15674–15679 (2013)
10. Lathrop, R.H.: The protein threading problem with sequence amino acid interaction preferences is NP-complete. Protein Eng. **7**(9), 1059–1068 (1994)
11. Boyd, S., et al.: Distributed optimization and statistical learning via the alternating direction method of multipliers. Found. Trends® Mach. Learn. **3**(1), 1–122 (2011)
12. Peng, J., Bo, L., Xu, J.: Conditional neural fields. In: Advances in Neural Information Processing Systems (2009)
13. Ma, J., et al.: Protein threading using context-specific alignment potential. Bioinformatics **29** (13), i257–i265 (2013)
14. Ma, J., et al.: A conditional neural fields model for protein threading. Bioinformatics **28**(12), i59–i66 (2012)
15. Wang, S., et al.: Protein structure alignment beyond spatial proximity, vol. 3. Science Report (2013)
16. Henikoff, S., Henikoff, J.G.: Amino acid substitution matrices from protein blocks. Proc. Natl. Acad. Sci. **89**(22), 10915–10919 (1992)
17. Prlić, A., Domingues, F.S., Sippl, M.J.: Structure-derived substitution matrices for alignment of distantly related sequences. Protein Eng. **13**(8), 545–550 (2000)
18. Tan, Y.H., Huang, H., Kihara, D.: Statistical potential-based amino acid similarity matrices for aligning distantly related protein sequences. Proteins Struct. Funct. Bioinf. **64**(3), 587–600 (2006)
19. Shen, M.Y., Sali, A.: Statistical potential for assessment and prediction of protein structures. Protein Sci. **15**(11), 2507–2524 (2006)
20. Zhang, J., Zhang, Y.: A novel side-chain orientation dependent potential derived from random-walk reference state for protein fold selection and structure prediction. PLoS ONE **5** (10), e15386 (2010)

21. Zhou, H., Zhou, Y.: Distance-scaled, finite ideal-gas reference state improves structure-derived potentials of mean force for structure selection and stability prediction. Protein Sci. **11**(11), 2714–2726 (2002)
22. Kabsch, W., Sander, C.: Dictionary of protein secondary structure: pattern recognition of hydrogen-bonded and geometrical features. Biopolymers **22**(12), 2577–2637 (1983)

Chapter 3
Software

Abstract This chapter introduces the software implementing the ideas presented in the Method chapter. It covers installation, usage and result interpretation of the software.

Keywords MRFalign · MRFsearch · P-value

MRFpred is a protein homology detection program that implements the above-described method. It is freely available at http://raptorx.uchicago.edu/download/.

3.1 Overview of Program

The MRFpred package (X86_64 Linux version with OpenMPI installed) consists of the following three main programs:

buildFeature Generating features for a given protein sequence.

MRFalign Aligning two protein sequences by aligning their corresponding MRFs. The program will take two protein feature files as input and output their alignment and some details such as alignment score.

MRFsearch Searching a given protein database for a query protein sequence to find its homologs. This program uses MRFalign to build alignments and then rank all the proteins in the database by alignment score.

You can simply run a program without arguments to obtain the help information.

3.2 Software Download

The whole MRFpred package includes the following files,

1. MRFsearch_v1.1.tar.gz ($\sim$100M): the package contains all the programs.
2. TGT_BC40.tar.gz ($\sim$900M): a protein database for homology search, in which any two proteins share <40 % sequence identity.

J. Xu et al., *Protein Homology Detection Through Alignment*
of Markov Random Fields, SpringerBriefs in Computer Science,
DOI 10.1007/978-3-319-14914-1_3

3. MRFsearch_Edge_Potential.tar.gz ($\sim$1G): this package contains pairwise residue correlation information for the proteins in the database.
4. nr70.tar.gz ($\sim$1.6G) and nr90.tar.gz ($\sim$2.3G): the formatted NR70 and NR90 protein sequence databases.
5. MRFsearch_Database_List.tar.gz ($\sim$1M): the package contains some database lists. The database lists will be updated weekly.

3.3 Feature Files

For each protein there are two feature files. One is a *.tgt file that contains some basic information about this protein and some predicted local features such as secondary structure. Notice that not all the features are use in MRFsearch. Below are some details.

Sequence information	Primal sequence	
Profile information	NEFF value	The average entropy of the protein sequence profile
	PSSM matrix	Position-specific score matrix generated by PSI-blast with 5 iterations and E-value 0.001
	PSFM matrix	Position-specific frequency matrix generated by PSI-blast with 5 iterations and E-value 0.001
	HMM matrix	The Hidden Markov Model generated by buildali2.pl in the HHpred package
Structure information	3-class secondary structure	The predicted 3-class secondary structure types by PSIPRED [1]
	8-class secondary structure	The predicted 8-class secondary structure types by RaptorX-SS8 [2]
	3-class solvent accessibility	The predicted 3-state solvent accessibility, which is discretized into three equal-frequency states: buried, intermediate and exposed
	Disorder prediction	The predicted disorder information by DISOPRED [3]

The other feature file contains pairwise residue correlation information produced by the EPAD package. For a protein of length N, it contains $N(N - 6)/2$ rows, where each rows contains 9 numbers representing the probability of 13 distance bins (3–4, 4–5, 5–6, ..., 14–15, and >15 Å).

3.4 MRFsearch Ranking File

After searching the database a ranking file will be generated as shown in Fig. 3.1. The first line contains the query protein name. The second line shows the query protein sequence. The third line is the query protein length. The NEFF (number of effective sequence homologs) in the fourth line is the average Shannon "Sequence Entropy" for a PSI-BLAST sequence profile. NEFF is the average number of amino acid (AA) substitutions across all residues of a protein, ranging from 1 to 20 (i.e., the number of AA types). NEFF at one residue is calculated by $\exp(-\sum_k p_k \ln p_k)$ where p_k is the probability for the kth AA type), and NEFF for the whole protein is the average across all residues. Generally speaking, NEFF is used to quantify the homologous information content available for a given protein. The larger the NEFF value, the more homologous information its profile contains. The fifth line contains the number of proteins searched by MRFsearch.

The meaning of each column is explained as follows.

Column 1 'No'	Ranking of all the searched proteins.
Column 2 'Proteins'	Name of the protein (PDB ID or SCOP protein name) in the databases.
Column 3 'P-value'	The P-value of the alignment. The smaller, the better.
Column 4 'Score'	The alignment raw score between the query and subject proteins.
Column 5 'Node'	The accumulative node alignment potential.

```
Query Name = 54180

Query Sequence = MLSVNTIKNTLLAAVLVSVPATAQVSGNGHPNLIVTEQDVANIAASWESYDAYAEQLNADKTNLDAFMAEGVVVPMPKDAGGGYTHEQHKRNYKAIRNAGFLYQVTGDEKYLTFAKDLLLAYAKMY
PSLGEHPNRKEQSPGRLFWQSLNEAVWLVYSIQGYDAIIDGLAAEEKQEIESGVFLPMAKFLSVESPETFNKIHNLGTWAVAAVGMTGYVLGNDELVEISLMGLDKTGKAGFMKQLDKLFSPDGYYTEGPYYQRYALMPFIWF
AKAIETNEPERKIFEYRNNILLKAVYTTIDLSYAGYFFPINDALKDKGIDTVELVHALAIVYSITGDNTLLDIAQEQGRISLTGDGLKVAKAVGEGLTQPYNYRSILLGDGADGDQGALSIHRLGEGHNHMALVAKNTSQGMG
HGHFDKLNWLLYDNGNEIVTDYGAARYLNVEAKYGGHYLAENNTWAKQTIAHNTLVVNEQSHFYGDVTTADLHHPEVLSFYSGEDYQLSSAKEANAYDGVEFVRSMLLVNVPSLEHPIVVDVLNVSADKASTFDLPLYFNGQI
IDFSFKVKDNKNVMKMLGKRNGYQHLWLRNTAPVGDASERATWILDDRFYSYAFVTSTPSKKQNVLIAELGANDPNYNLRQQQVLIRRVEKAKQASFVSVLEPHGKYDGSLETTSGAYSNVKSVKHVSENGKDVVVVDLKDGS
NVVVALSYNANSEQVHKVNAGEEAIEWKGFSSVVVRRK

Query Length = 736

Query NEFF = 8.1

Searched Templates = 8353
----------------------------------------------------------------------
Date Sun Aug 31 18:57:00 2014

No  Template  Pvalue    Score   Node    Edge    qRange    tRange   tLength  Cols   #tGaps  #qGaps  #seqID

1   d1itwa_   0.002914  294.4   81.48   212.9   3-736     35-722   740      646    88      42      75

2   d1jb0b_   0.006884  264.1   68.59   195.5   138-736   1-674    739      549    50      125     83

3   d1u0fa_   0.009197  253.9   67.67   186.2   45-543    1-556    556      472    27      84      67

4   d1oaod_   0.01077   248.3   58.3    190     265-736   17-574   728      456    16      102     57

5   d1wdpa1   0.01081   248.2   76.29   171.9   26-502    3-490    490      447    30      41      57

6   d1n7va_   0.01113   247.2   84.76   162.4   127-660   62-555   555      451    83      43      56

7   d1orva1   0.01126   246.8   74.31   172.4   303-717   1-470    470      392    23      78      42

8   d1xfda1   0.01266   242.6   75.57   167.1   308-718   43-465   465      386    25      37      40

9   d1gqia1   0.01528   236     65.23   170.7   240-666   14-561   561      410    17      138     54

10  d1acca_   0.01662   233     59.79   173.2   207-736   1-475    722      460    70      15      52
```

Fig. 3.1 An example ranking file generated by MRFsearch

Column 6 'Edge'   The accumulative edge alignment potential.
Column 7 'qRange'   Range of aligned region in the query protein.
Column 8 'tRange'   Range of aligned region in the subject protein.
Column 9 'tLength'   The length of the subject protein.
Column 10 'Cols'   The number of aligned positions in the pairwise alignment.
Column 11 '#tGaps'   The number of gaps in the subject protein (Insertions).
Column 12 '#qGaps'   The number of gaps in the query protein (Deletions).
Column 13 '#seqID'   The number of identical residues in the alignment.

3.5 Interpreting P-Value

In the ranking file, P-value can be interpreted as a confidence score indicating the relative quality of the top-ranked proteins and (corresponding) alignments. To calculate the P-value, we employs a set of reference proteins (in "databases/ CAL_TGT)"), which consists of $\sim 1,800$ single-domain proteins belonging to different SCOP folds. Given a query protein, we first align it to this reference protein database and then estimate an extreme value distribution from the $\sim 1,800$ alignment scores. Based upon this distribution, we calculate the P-value of each alignment when aligning the query protein to the subject protein database. The P-value actually measures the likelihood of each subject protein being homologous to the query protein by comparing it to the reference proteins.

To see the relationship between the P-value and the closeness of the first-ranked protein by MRFsearch to a query protein, we conduct an experiment on the 368 CAMEO target proteins. For each CAMEO target, the first-ranked protein in the database is treated as the homolog of this target. To measure the quality of an alignment, we use un-normalized Global Distance Test (GDT). GDT has been employed as an official measure of a protein model quality by CASP for many years. When applied to alignments, uGDT can be interpreted as the number of correctly-aligned positions in an alignment, but weighted by alignment quality at each position. We say one alignment is good when its uGDT is larger than 50. We use 50 as a cutoff because that many proteins similar at only the fold level have uGDT around 50. Figure 3.2 shows the relationship between P-value and uGDT on the 368 CAMEO targets. Figure 3.3 is a zoom-in graph of Fig. 3.2, showing relationship between P-value and uGDT on the 132 CAMEO targets with $-\log(\text{P-value}) < 20$. As shown in Fig. 3.3, when P-value is small (i.e. $<10\text{e}-10$), most alignments have uGDT greater than or equal to 50. That is, when P-value is less than $10\text{e}-10$, the first-ranked protein is very likely to share a similar fold as the query protein. When P-value is between $10\text{e}-5$ and $10\text{e}-10$, more than half of the alignments have uGDT > 50.

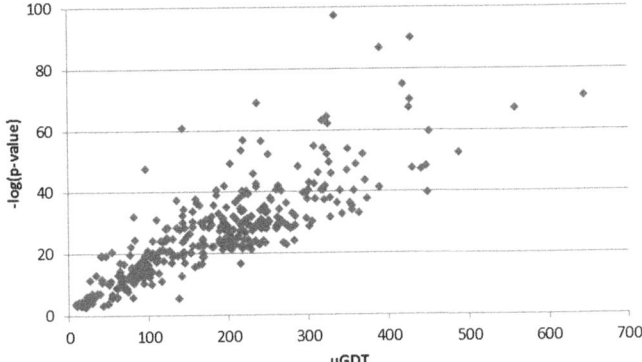

Fig. 3.2 The relationship between P-value and the alignment quality of the first-ranked homologs on the 368 CAMEO targets. The x-axis is the alignment quality measured by uGDT and the y-axis is −log(P-value)

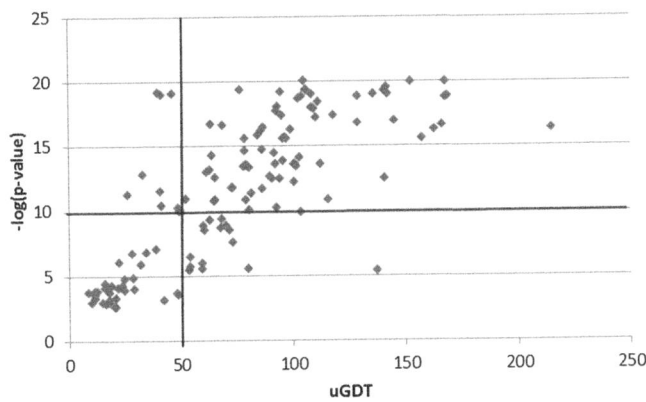

Fig. 3.3 The relationship between P-value and the alignment quality of the first-ranked homologs on the 132 CAMEO targets with −log(P-value) < 20. The x-axis is the alignment quality measured by uGDT and the y-axis is −log(P-value)

3.6 Interpreting a Pairwise Alignment

The last section in the ranking file contains the alignments of the query protein to the top-ranked subject proteins. Figure 3.4 shows an example alignment between the query and subject protein. The alignment consists of one or more blocks with the following lines: the first line 'T T0644' is the subject protein name and the third line 'S T0645' is the query protein name, each followed by its sequence. The line in the middle shows the alignment score at each aligned positions. The higher the score, the better the alignment is. The symbols used here have the following meanings.

```
T T0644          1 DDDTGYLPPSQAIQDALKKLYPNATAIKWEQKGVYY   36 (141)
                   .~*|*~*************~*~*************
S T0645        463 DDAKKDPEQYKNLCTALGGADNGGTRLWWDTGKNNF  498 (498)
```

Fig. 3.4 An example alignment in the ranking file and the alignment detail file

'|'   Alignment score above 13.
'+'   Alignment score between 10 and 12.
'*'   Alignment score between 6 and 9.
'~'   Alignment score between 2 and 5.
'.'   Alignment score between −1 and 1.
'−'   Alignment score between −4 and −2.
'='   Alignment score below −5.

References

1. Jones, D.T.: Protein secondary structure prediction based on position-specific scoring matrices. J. Mol. Biol. **292**(2), 195–202 (1999)
2. Wang, Z., et al.: Protein 8-class secondary structure prediction using conditional neural fields. Proteomics **11**(19), 3786–3792 (2011)
3. Jones, D.T., Ward, J.J.: Prediction of disordered regions in proteins from position specific score matrices. Proteins Struct. Funct. Bioinf. **53**(S6), 573–578 (2003)

Chapter 4
Experiments and Results

Abstract This chapter describes the experimental results of the MRF-based method for homology detection and fold recognition including alignment accuracy, success rate, running time and contribution of some important features. This chapter also compares the MRF-based method with currently popular PSSM- and HMM-based methods such as HHpred, HHblits and FFAS, in terms of alignment accuracy and success rate of homology detection and fold recognition.

Keywords Alignment accuracy · Homology detection success rate · Fold recognition rate · HHpred · HHblits · FFAS

4.1 Training and Validation Data

To train the node alignment potential, we constructed the training and validation data from SCOP70, in which any two proteins share <70 % sequence identity. In total we use a set of 1,400 protein pairs as the training and validation data, which covers 458 SCOP fold [1–3]. The sequence identity of all the training and validation protein pairs is uniformly distributed between 20 and 70 %. Further, two proteins in a pair are similar at superfamily or fold level. A training or validation protein has less than 400 residues and contains less than 10 % of residues without 3D coordinates. The reference alignment for a protein pair is generated by a structure alignment tool DeepAlign. Each reference alignment has fewer than 50 middle gap positions and the number of terminal gaps is less than 20 % of the alignment length. Five-fold cross validation is used to determine the hyper-parameters in our machine learning model. In particular, each time we choose 1,000 out of the 1,400 protein pairs as the training data and the remaining 400 pairs as the validation data such that there is no fold-level redundancy between the training and validation data.

© The Author(s) 2015
J. Xu et al., *Protein Homology Detection Through Alignment of Markov Random Fields*, SpringerBriefs in Computer Science,
DOI 10.1007/978-3-319-14914-1_4

4.2 Test Data

The data used to test alignment accuracy has no fold-level overlap with the training and validation data. In particular, we use the following three datasets to test the alignment accuracy, which are subsets of the test data used in [4] to benchmark protein modeling methods.

- Set3.6K: a set of 3,617 non-redundant protein pairs. Two proteins in a pair share <40 % sequence identity and have small length difference. By "non-redundant" we mean that in any two protein pairs, there are at least two proteins (one from each pair) sharing less than 25 % sequence identity.
- Set2.6K: a set of 2,633 non-redundant protein pairs. Two proteins in a pair share <25 % sequence identity and have length difference larger than 30 %. This set is mainly used to test an alignment method in handling with domain boundary.
- Set60K: a very large set of 60,929 protein pairs, in most of which two proteins share less than 40 % sequence identity. Meanwhile, 846, 40,902, and 19,181 pairs are similar at the SCOP family, superfamily and fold level, respectively, and 151, 2,691 and 2,218 pairs consist of only mainly beta proteins, respectively.

We use three benchmarks SCOP20, SCOP40 and SCOP80 to test the success rate of remote homology detection and fold recognition. These benchmarks were used by Söding group to study context-specific mutation score [4]. They are constructed by filtering the SCOP database with a maximum sequence identity of 20, 40 and 80 %, respectively. In total they have 4,884, 7,088, and 9,867 proteins, respectively, and 1,281, 1,806, and 2,734 mainly beta proteins, respectively.

For a protein in the first three datasets, we run PSI-BLAST with 5 iterations to find close sequence homologs and the build multiple sequence alignment (MSA). The MSA files for the three SCOP benchmarks are downloaded from the HHpred website (ftp://toolkit.genzentrum.lmu.de/pub/). Pseudo-counts are employed especially for proteins with very few close sequence homologs.

Programs to compare To evaluate alignment accuracy, we compare our method, denoted as MRFalign, with sequence-HMM alignment method HMMER [5] and HMM-HMM alignment method HHalign [6]. HHMER is run with a default E-value threshold (10.0). HHalign is run with the option "-mact 0.1". To evaluate the performance of remote homology detection, we compare MRFalign, with PSSM-PSSM based method FFAS [7], sequence-HMM based method hmmscan (comparison) and HMM-HMM based methods HHsearch and HHblits [8]. Meanwhile, HHsearch and hmmscan use HHalign and HMMER, respectively, for protein alignment.

Evaluation criteria Three performance metrics are employed including reference-dependent alignment recall and precision, and success rate of homology detection and fold recognition. Alignment recall is the fraction of align-able residues in a reference alignment that are aligned by an algorithm. Alignment precision is defined as the fraction of aligned positions that are correct according to a reference alignment. Since reference alignments are used to judge if one residue is correctly aligned or align-able, their quality is critical. To reduce bias, we use three very different structure alignment tools TM-align [9], Matt [10], and DeepAlign [11] to generate reference alignments.

4.3 Reference-Dependent Alignment Recall

As shown in Tables 4.1 and 4.2, our method MRFalign exceeds all the others regardless of the reference alignments on both dataset Set3.6K and Set2.6K. When only exact matches are considered as correct, MRFalign outperforms HHalign by ~ 10 % on both datasets, and HHMER by ~ 23 and ~ 24 % on Set3.6K and Set2.6K, respectively. If 4-position off exact match is allowed in calculating alignment recall, MRFalign outperforms HHalign by ~ 11 % on both datasets, and HHMER by ~ 25 and ~ 33 %, respectively.

On the very large set Set60K, as shown in Tables 4.3, 4.4 and 4.5, MRFalign outperforms the other two at each SCOP classification level regardless of the employed reference alignments. MRFalign is only slightly better than HHalign at the family level, which is not surprising since it is easy to align two closely-related proteins. At the superfamily level, MRFalign outperforms HHalign and HMMER

Table 4.1 Reference-dependent alignment recall on Set3.6K

	TMalign		Matt		DeepAlign	
	Exact match (%)	4-offset (%)	Exact match (%)	4-offset (%)	Exact match (%)	4-offset (%)
HMMER	22.9	26.5	24.1	27.4	25.5	28.1
HHalign	36.3	39.1	37.0	42.1	38.4	42.8
MRFalign	47.4	51.0	47.5	52.6	49.2	53.5

Three structure alignment tools (TMalign, Matt and DeepAlign) are used to generate reference alignments. "4-offset" means that 4-position off the exact match is allowed. The bold indicates the best results

Table 4.2 Reference-dependent alignment recall on Set2.6K

	TMalign		Matt		DeepAlign	
	Exact match (%)	4-offset (%)	Exact match (%)	4-offset (%)	Exact match (%)	4-offset (%)
HMMER	36.5	42.6	38.6	44.0	40.4	45.0
HHalign	62.5	66.1	63.2	66.2	64.0	66.7
MRFalign	72.8	76.2	73.5	76.7	74.2	77.8

See Table 4.1 for explanation

Table 4.3 Reference-dependent alignment recall on the large benchmark Set60K

	HMMER (%)	HHalign (%)	MRFalign (%)
Family	57.4	69.2	71.0
Superfamily	31.2	42.0	48.1
Fold	1.3	7.0	14.2
Family (beta)	60.9	69.9	73.1
Superfamily (beta)	35.0	47.2	52.1
Fold (beta)	2.5	8.3	17.3

Only exact matches are considered as correct in evaluating alignment quality. The test protein pairs are divided into 3 groups based upon the SCOP classification. The bold indicates the best results. The structure alignments generated by TMalign are used as reference alignments

Table 4.4 Reference-dependent alignment recall on the large benchmark Set60K

	HMMER (%)	HHalign (%)	MRFalign (%)
Family	59.1	70.5	74.5
Superfamily	32.3	42.4	51.7
Fold	1.6	8.0	15.5
Family (beta)	64.0	75.1	78.4
Superfamily (beta)	37.0	50.2	55.8
Fold (beta)	3.0	9.1	17.1

The structure alignments generated by Matt are used as reference alignments. See Table 4.3 for more explanation

Table 4.5 Reference-dependent alignment recall on the large benchmark Set60K

	HMMER (%)	HHalign (%)	MRFalign (%)
Family	63.2	72.6	75.5
Superfamily	32.8	49.4	55.6
Fold	2.0	8.7	18.4
Family (beta)	68.4	79.0	82.9
Superfamily (beta)	39.1	52.9	60.7
Fold (beta)	4.0	10.1	21.8

The structure alignments generated by DeepAlign are used as reference alignments. See Table 4.3 for more explanation

by ~ 6 and ~ 18 %, respectively. At the fold level, MRFalign outperforms HHalign and HHMER by ~ 7 and ~ 14 %, respectively.

Alignment recall for beta proteins As shown in Tables 4.3, 4.4 and 4.5, our method MRFalign outperforms HHalign and HMMER by ~ 3 and ~ 12 %, respectively, at the family level; ~ 7 and ~ 19 %, respectively, at the superfamily level; and ~ 10 and ~ 16 %, respectively, at the fold level, regardless of reference alignments.

4.4 Reference-Dependent Alignment Precision

As shown in Tables 4.6 and 4.7, our method MRFalign exceeds all the others regardless of the reference alignments on both data sets Set3.6K and Set2.6K. When only exact matches are considered as correct, MRFalign outperforms HHalign by ∼8 and ∼5 % on Set3.6K and Set2.6K, respectively, and HHMER by ∼15 and ∼13 %, respectively. If 4-position off the exact match is allowed in calculating alignment precision, MRFalign outperforms HHalign by ∼8 and ∼9 %, and HHMER by ∼14 and ∼18 % on Set3.6K and Set2.6K, respectively (Tables 4.9 and 4.10).

Table 4.6 Reference-dependent alignment precision on Set3.6K

	TMalign		Matt		DeepAlign	
	Exact match (%)	4-offset (%)	Exact match (%)	4-offset (%)	Exact match (%)	4-offset (%)
HMMER	29.3	34.1	29.6	34.7	31.5	35.6
HHalign	35.9	39.4	36.2	39.4	37.2	41.7
MRFalign	43.2	47.4	44.1	48.5	46.1	50.4

Three structure alignment tools (TMalign, Matt and DeepAlign) are used to generate reference alignments. "4-offset" means that 4-position off the exact match is allowed. The bold indicates the best results

Table 4.7 Reference-dependent alignment precision on Set2.6K

	TMalign		Matt		DeepAlign	
	Exact match (%)	4-offset (%)	Exact match (%)	4-offset (%)	Exact match (%)	4-offset (%)
HMMER	48.0	50.1	48.2	50.3	51.4	54.8
HHalign	57.1	59.9	57.3	60.0	58.3	61.4
MRFalign	62.5	69.1	62.7	69.6	63.2	70.0

See Table 4.6 for more explanation

Table 4.8 Reference-dependent alignment precision on the large benchmark Set60K

	HMMER (%)	HHalign (%)	MRFalign (%)
Family	63.1	63.9	67.3
Superfamily	38.7	39.5	42.8
Fold	4.2	7.4	11.5
Family (beta)	66.4	65.8	69.5
Superfamily (beta)	44.2	44.9	48.8
Fold (beta)	6.1	9.3	14.1

Only exact matches are considered correct in evaluating alignment quality. The protein pairs are divided into 3 groups based upon the SCOP classification. The bold indicates the best results. The structure alignment generated by TMalign are used as reference alignments

Table 4.9 Reference-dependent alignment precision on the large benchmark Set60K

	HMMER (%)	HHalign (%)	MRFalign (%)
Family	64.3	65.4	68.0
Superfamily	40.5	41.3	44.9
Fold	4.7	8.0	12.3
Family (beta)	67.4	68.1	72.3
Superfamily (beta)	45.4	46.2	49.4
Fold (beta)	6.7	9.2	14.5

The structure alignments generated by Matt are used as reference alignments. See Table 4.8 for more explanation

Table 4.10 Reference-dependent alignment precision on the large benchmark Set60K

	HMMER (%)	HHalign (%)	MRFalign (%)
Family	68.4	69.2	71.4
Superfamily	43.2	44.3	48.7
Fold	5.4	8.2	14.5
Family (beta)	70.8	72.4	77.9
Superfamily (beta)	46.6	48.4	53.7
Fold (beta)	7.9	8.6	17.8

The structure alignments generated by DeepAlign are used as reference alignments. See Table 4.8 for more explanation

On the very large set Set60K, as shown in Table 4.6, MRFalign outperforms the other two at each SCOP classification level regardless of the reference alignments used. At the family level, MRFalign outperforms HHalign and HMMER by ∼3 and ∼4 %, respectively. At the superfamily level, our method outperforms HHalign and HMMER by ∼4 and ∼5 %, respectively. At the fold level, MRFalign outperforms HHalign and HHMER by ∼5 and ∼8 %, respectively.

4.5 Success Rate of Homology Detection and Fold Recognition

To evaluate the success rate of homology detection and fold recognition, we employ three benchmarks SCOP20, SCOP40 and SCOP80 introduced in [4]. For each protein sequence in one benchmark, we treat it as a query, align it to all the other proteins in the same benchmark and then examine if the query is similar to those with the best alignment scores or not. We also tested the performance of on these benchmarks hmmscan [5], FFAS [7], HHsearch [6] and HHblits [8], all of which are run with default options. The success rate is measured at the superfamily and fold levels, respectively. When evaluating the success rate at the superfamily (fold) level, we exclude those proteins similar to the query at least at the family (superfamily) level. For each query protein, we examine the top 1-, 5- and 10-ranked proteins, respectively.

As shown in Table 4.11, tested on SCOP20, SCOP40 and SCOP80 at the superfamily level, our method MRFalign succeeds on $\sim$6, $\sim$4 and $\sim$4 % more query proteins than HHsearch, respectively, when only the first-ranked proteins are considered. As shown in Table 4.11, at the fold level, MRFalign succeeds on $\sim$11, $\sim$11 and $\sim$12 % more proteins than HHsearch, respectively, when only the first-ranked proteins are evaluated. At the superfamily level, SCOP20 is more challenging than the other two benchmarks because it contains fewer proteins similar at this level. Nevertheless, at the fold level, SCOP80 is slightly more challenging than the other two benchmarks maybe because it contains many more irrelevant proteins and thus, the chance of ranking false positives at top is higher.

Similar to alignment accuracy, MRFalign for homology detection also has a larger advantage on the beta proteins. In particular, as shown in Table 4.13, tested on SCOP20, SCOP40 and SCOP80 at the superfamily level, MRFalign succeeds on $\sim$7, $\sim$5 and $\sim$7 % more proteins than HHsearch, respectively, when only the first-ranked proteins are evaluated. As shown in Table 4.12, at the fold level, MRFalign succeeds on $\sim$13, $\sim$16 and $\sim$17 % more proteins than HHsearch, respectively, when only the first-ranked proteins are evaluated. Note that in this experiment, only the query proteins are mainly-beta proteins, the subject proteins may be of any types. If we restrict the subject proteins to only beta proteins, the success rate increases further due to the reduction of false positives (Table 4.14).

Table 4.11 Homology detection success rate (%) at the superfamily level on three benchmarks SCOP20, SCOP40 and SCOP80

	SCOP20			SCOP40			SCOP80		
	Top1	Top5	Top10	Top1	Top5	Top10	Top1	Top5	Top10
Hmmscan	35.2	36.5	36.5	40.2	41.7	41.8	43.9	45.2	45.3
FFAS	48.6	54.4	55.6	52.1	56.3	57.1	49.8	53.0	53.7
HHsearch	51.6	57.3	59.2	55.8	60.8	62.4	56.1	60.1	61.8
HHblits	51.9	56.3	57.5	56.0	59.8	60.9	59.2	62.5	63.3
MRFalign	58.2	61.7	63.4	59.3	63.6	65.8	60.4	64.7	66.1

Table 4.12 Homology detection success rate (%) at the fold level on three benchmarks SCOP20, SCOP40 and SCOP80

	SCOP20			SCOP40			SCOP80		
	Top1	Top5	Top10	Top1	Top5	Top10	Top1	Top5	Top10
Hmmscan	5.2	6.1	6.1	6.2	6.9	6.9	5.9	6.5	6.6
FFAS	13.1	18.7	20.0	10.4	14.5	15.4	9.1	11.9	12.6
HHsearch	16.3	24.7	28.6	17.6	25.3	29.1	15.4	21.9	25.0
HHblits	17.4	25.2	27.2	19.1	26.0	28.2	18.4	25.0	27.0
MRFalign	27.2	36.8	41.2	28.3	37.9	42.4	27.0	38.1	41.6

Table 4.13 Homology detection success rate (%) for mainly beta proteins at the superfamily level on three benchmarks SCOP20, SCOP40 and SCOP80

	SCOP20			SCOP40			SCOP80		
	Top1	Top5	Top10	Top1	Top5	Top10	Top1	Top5	Top10
Hmmscan	29.1	29.4	29.4	34.7	35.1	35.1	43.7	44.0	44.1
FFAS	43.6	49.9	51.9	48.2	52.4	53.5	43.7	46.3	47.2
HHsearch	48.2	54.6	56.9	52.0	56.9	59.1	47.7	51.8	53.7
HHblits	47.5	52.1	53.7	51.4	54.8	56.6	52.9	54.6	57.8
MRFalign	55.4	61.7	65.9	57.3	63.5	66.8	54.2	59.7	64.2

Table 4.14 Homology detection success rate (%) for mainly beta proteins at the fold level on three benchmarks SCOP20, SCOP40 and SCOP80

	SCOP20			SCOP40			SCOP80		
	Top1	Top5	Top10	Top1	Top5	Top10	Top1	Top5	Top10
Hmmscan	6.9	7.6	7.6	8.0	8.6	8.6	7.0	7.4	7.4
FFAS	22.7	30.1	31.8	15.2	20.4	21.7	11.8	15.3	16.1
HHsearch	24.4	34.7	38.8	26.8	37.7	41.6	19.1	26.8	29.5
HHblits	24.1	33.3	34.8	26.9	35.3	37.1	24.7	34.1	35.5
MRFalign	37.4	55.0	61.4	42.5	51.0	54.6	36.4	48.0	55.9

4.6 Contribution of Edge Alignment Potential and Mutual Information

To evaluate the contribution of the edge alignment potential, we calculate the alignment recall improvement resulting from using the edge alignment potential on two benchmarks Set3.6K and Set2.6K. As shown in Tables 4.15 and 4.16, the edge alignment potential can improve alignment recall by 3.4 and 3.7 %, respectively. When mutual information is used to measure residue co-evolution, alignment recall can be further improved by 1.1 and 1.9 % on these two sets, respectively. Mutual

Table 4.15 Contribution of the edge alignment potential and mutual information (MI), measured by alignment recall improvement on two benchmarks Set3.6K and Set2.6K

	Set3.6K		Set2.6K	
	Exact match (%)	4-offset (%)	Exact match (%)	4-offset (%)
Only node potential	44.7	48.6	68.6	71.8
Node + edge potential, no MI	48.1	52.2	72.3	75.2
Node + edge potential with MI	49.2	53.5	74.2	77.8

The structure alignments generated by DeepAlign are used as reference alignments

Table 4.16 Contribution of the edge alignment potential and mutual information (MI), measured by alignment recall improvement on proteins with at least 256 non-redundant sequence homologs in two benchmarks Set3.6K and Set2.6K

	391 pairs in Set3.6K		509 pairs in Set2.6K	
	Exact match (%)	4-offset (%)	Exact match (%)	4-offset (%)
Only node potential	59.5	63.4	71.3	75.8
Node + edge potential, no MI	62.1	66.7	73.5	78.1
Node + edge potential with MI	65.2	69.8	76.6	81.0

The structure alignments generated by DeepAlign are used as reference alignments

information is mainly useful for proteins with many sequence homologs since it is close to 0 for proteins with few sequence homologs. As shown in Tables 4.15 and 4.16, if only the proteins with at least 256 non-redundant sequence homologs are considered, the improvement resulting from mutual information is ~ 3 %.

4.7   Running Time

Figure 4.1 shows the running time of MRFalign with respect to protein length. As a control, we also show the running time of the Viterbi algorithm, which is used by our ADMM algorithm to generate alignment at each iteration. As shown in this figure, MRFalign is no more than 10 times slower than the Viterbi algorithm. To speed up homology detection, we may use the Viterbi algorithm to perform an initial search without considering edge alignment potential, and keep only top 10 % of proteins for further examination. Then we run MRFalign to search for homologs from the top 10 %. Therefore, although MRFalign may be slow compared to the Viterbi algorithm, empirically we can do homology search only slightly slower than the Viterbi algorithm.

4.8   Is Our MRFalign Method Overtrained?

We conducted two experiments to show that MRFalign is not overtrained. In the first experiment, we used 36 CASP10 hard targets as the test data. Since our training set was built before CASP10 started, we can believe that there is no redundancy between the CASP10 hard targets and our training data. Using MRF-align and HHpred, respectively, we search each of these 36 test targets against

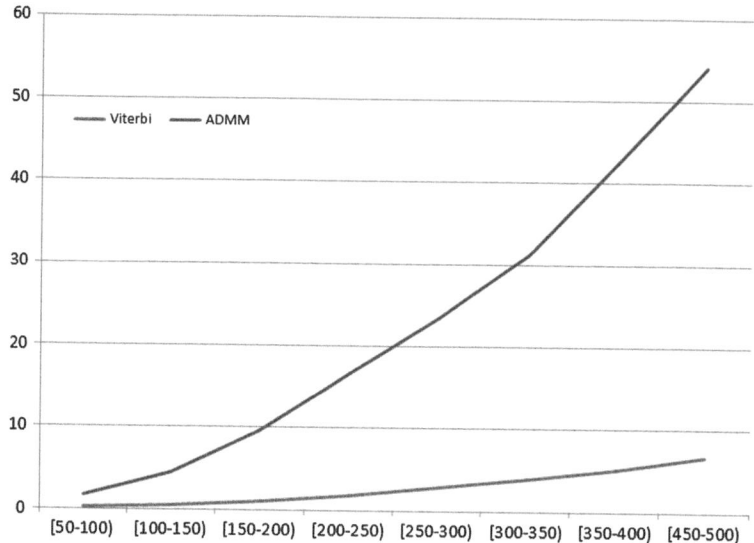

Fig. 4.1 Running time of the Viterbi algorithm and MRFalign (which uses the ADMM algorithm). The X-axis is the geometric mean of the lengths of two proteins under alignment. The Y-axis is the running time in seconds

PDB25 to find the best match. Since PDB25 does not contain proteins very similar to many of the test targets, we built a 3D model using MODELLER from the alignment between a test target and its best match and then measure the quality of the model. As shown in Fig. 4.2, MRFalign yields much better 3D models than HHsearch for most of the targets. This implies that MRFalign can generalize well to the test data not similar to the training data.

In the second experiment, we divide the proteins in SCOP40 into three subsets according their similarity with all the training data. We measure the similarity of one test protein with all the training data by its best BLAST E-value. We used two values 1e−2 and 1e−35 as the E-value cutoff so that the three subsets have roughly the same size. As shown in Table 4.17, the advantage of MRFalign in remote homology detection over HHpred is roughly same across the three subsets. Since HHpred is an unsupervised algorithm, this implies that the performance of MRF-align is not correlated to the similarity between the test and the training data. Therefore, it is unlikely that MRFalign is overfit by the training data.

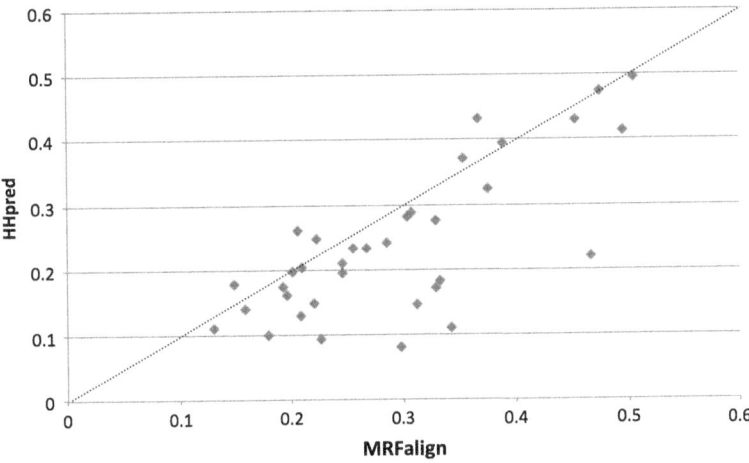

Fig. 4.2 The model quality, measured by TM-score, of our method and HHpred for the 36 CASP10 hard targets. Each point represents two models generated by our method (X-axis) and HHpred (Y-axis), respectively. TM-score ranges from 0 to 1 with 0 indicating the worst quality and 1 the highest

Table 4.17 Fold recognition rate of MRFalign on SCOP40, with respect to the similarity (measured by E-value) between the training and test data

	E < 1e−35			1e−35 < E < 1e−2			E > 1e−2		
	Top1	Top5	Top10	Top1	Top5	Top10	Top1	Top5	Top10
Hmmscan	5.0	5.6	5.6	7.3	7.9	7.9	6.4	7.3	7.4
FFAS	10.3	14.5	15.8	9.7	12.9	13.5	11.6	16.5	17.5
HHsearch	16.0	23.2	26.5	18.5	26.2	30.3	18.9	27.2	31.7
HHblits	16.9	23.1	25.5	20.8	27.4	28.9	20.2	28.3	31.1
MRFalign	25.5	35.9	39.4	29.7	39.5	43.3	29.4	39.0	43.6

The numbers are presented as percentage

References

1. Murzin, A.G.: SCOP: a structural classification of proteins database for the investigation of sequences and structures. J. Mol. Biol. **247**(4), 536–540 (2013)
2. Andreeva, A.: SCOP database in 2004: refinements integrate structure and sequence family data. Nucleic Acids Res. **32**(1), 226–229 (2004)
3. Andreeva, A.: Data growth and its impact on the SCOP database: new developments. Nucleic Acids Res. **36**(1), D419–D425 (2008)
4. Angermüller, C., Biegert, A., Söding, J.: Discriminative modelling of context-specific amino acid substitution probabilities. Bioinformatics **282**(4), 3240–3247 (2012)
5. Eddy, S.R.: HMMER: Profile Hidden Markov Models for Biological Sequence Analysis (2001)

6. Söding, J.: Protein homology detection by HMM-HMM comparison. Bioinformatics **21**(7), 951–960 (2005)
7. Rychlewski, L.: Comparison of sequence profiles. Strategies for structural predictions using sequence information. Protein Sci. **9**(2), 232–241 (2000)
8. Remmert, M.: HHblits: lightning-fast iterative protein sequence searching by HMM-HMM alignment. Nat. Methods **9**(2), 173–175 (2012)
9. Zhang, Y., Skolnick, J.: TM-align: a protein structure alignment algorithm based on the TM-score. Nucleic Acids Res. **33**(7), 2302–2309 (2005)
10. Menke, M., Berger, B., Cowen, L.: Matt: local flexibility aids protein multiple structure alignment. PLoS Comput Biol **4**(1), e10 (2008)
11. Wang, S., et al.: Protein structure alignment beyond spatial proximity, vol. 3. Scientific Reports (2013)

Conclusion

This book has presented a new method for sequence-based protein homology detection that compares two proteins through alignment of two Markov Random Fields (MRFs), which model the multiple sequence alignment (MSA) of a protein set using an undirected general graph in a probabilistic way. The MRF representation is better than the extensively-used PSSM (position-specific scoring matrix) and HMM (Hidden Markov Model) representations in that the former can model long-range residue interactions while the latter two cannot. As such, MRF-based homology detection shall be much more sensitive than PSSM- and HMM-based methods. Our large-scale experimental tests show that MRF-MRF comparison can greatly improve alignment accuracy and remote homology detection over currently popular sequence-HMM, PSSM-PSSM, and HMM-HMM comparison methods. Our method also has a larger advantage over the others on mainly-beta proteins.

We build our MRF model from multiple sequence alignment (MSA) without using any native structures, so the accuracy of an MRF model depends on the accuracy of an MSA. Currently our MRF model is built upon the MSA generated by PSI-BLAST. In the future, we may explore better alignment methods for MSA building or even utilize a few solved structures to improve MSA. The accuracy of the MRF model parameter usually increases with respect to the number of non-redundant sequence homologs in the MSA. Along with more and more protein sequences are generated, very accurate MRFs will be available for more and more protein families and thus, their homologous relationship can be studied more accurately using MRFs.

An accurate scoring function is essential to MRF-MRF comparison. Although in this book we only present one scoring function, various scoring functions can be used to measure the similarity of two MRFs, just like quite a few scoring functions are developed to measure the similarity of two PSSMs or HMMs. It is computationally intractable to find the best alignment between two MRFs when long-range residue interaction is considered. This book presents an ADMM algorithm that can efficiently solve the MRF-MRF alignment problem to suboptimal. However, it is about 10 times slower than the dynamic programming algorithm for PSSM-PSSM alignment. Further tuning of this ADMM algorithm is needed for very large-scale homology detection on a laptop computer.

© The Author(s) 2015
J. Xu et al., *Protein Homology Detection Through Alignment of Markov Random Fields*, SpringerBriefs in Computer Science,
DOI 10.1007/978-3-319-14914-1

Acknowledgements

This work is supported by National Institutes of Health grant R01GM089753, NSF CAREER Award CCF-1149811 and Alfred P. Sloan Research Fellowship. The authors are also grateful to the University of Chicago Beagle team and TeraGrid for their support of computational resources. The funders had no role in study design, data collection and analysis, decision to publish, or preparation of the manuscript.

© The Author(s) 2015
J. Xu et al., *Protein Homology Detection Through Alignment*
of Markov Random Fields, SpringerBriefs in Computer Science,
DOI 10.1007/978-3-319-14914-1